Beihang Postgraduate Series

北京航空航天大学"研究生英文教材"系列丛书

Design, Fabrication and CAD for MEMS Devices

微机电器件设计、制造及计算机辅助设计

Guo Zhanshe

郭占社

北京航空航天大学出版社

Abstract

This book firstly introduces the pertinent fundamental theory, important material and fabrication process of micro-electromechanical systems. Based on these theories, the design rule and important engineering examples are described in detail. Then, many engineering applications for MEMS including the acceleration measurement, the angular speed measurement and the pressure measurement are introduced. Finally, finite element method is introduced in order to prove the correctness of the design. This engineering application of simulation includes the static and modal analysis of the beam, capacitance analysis, thermal-structure analysis of the device and fatigue analysis etc.

It can be selected as the reference to the postgraduates, undergraduates and pertinent engineering staff whose research directions are instrumentation science and technology, control science and engineering, mechanical engineering etc.

图书在版编目(CIP)数据

微机电器件设计、制造及计算机辅助设计 = Design, Fabrication and CAD for MEMS Devices：英文 / 郭占社编著. -- 北京：北京航空航天大学出版社，2016.5
ISBN 978-7-5124-2109-7

Ⅰ.①微… Ⅱ.①郭… Ⅲ.①微电子技术—高等学校—教材—英文 Ⅳ.①TN4

中国版本图书馆 CIP 数据核字(2016)第 081763 号

版权所有，侵权必究。

Design, Fabrication and CAD for MEMS Devices
微机电器件设计、制造及计算机辅助设计
Guo Zhanshe
郭占社
责任编辑　冯　颖
*
北京航空航天大学出版社出版发行
北京市海淀区学院路37号(邮编100191)　http://www.buaapress.com.cn
发行部电话：(010)82317024　传真：(010)82328026
读者信箱：bhpress@263.net　邮购电话：(010)82316936
北京时代华都印刷有限公司印装　各地书店经销
*
开本：787×960　1/16　印张：17.5　字数：392千字
2016年5月第1版　2016年5月第1次印刷　印数：1500册
ISBN 978-7-5124-2109-7　定价：65.00元

若本书有倒页、脱页、缺页等印装质量问题，请与本社发行部联系调换。联系电话：(010)82317024

Preface

MEMS (Micro-ElectroMechanical Systems) is an emerging field developed in recent decades. It's a frontier technology in twenty-first century based on IC (Integrated Circuits) technology. It mainly refers to a technology of design, processing, manufacturing, measurement and control based on silicon materials. It is a micro-system which can integrate mechanical components, optical systems, driving components and electronic control systems on a single chip. It is a new type of multi-disciplinary technique developed in recent years. It involves machinery, electronics, chemistry, physics, optics, biology, materials and other disciplines. It is one of the key research fields at present.

In this book, the basic theory, important materials, the fabrication process, the rule of MEMS design and many engineering examples are introduced in detail combining the author's research experience and results. Then, some kinds of typical MEMS measuring devices including the MEMS accelerometer, angular measurement sensors, pressure sensors, micro-torque measuring methods are introduced in detail combining the author's research experience. Finally, many engineering examples for the application of finite element method in MEMS are introduced.

Chapter 1 introduces the definition of MEMS in two aspects: the dimension and the fabrication process. Then, the development process and the application of CAD in MEMS are also introduced.

Chapter 2 introduces the detailed theory of MEMS components such as the single clamped beam, the double-enddamped beam, the comb finger actuator and the diagraph etc.

In Chapter 3, the mechanical characters and some important problems to be noticed for some kinds of typical materials are mentioned.

Chapter 4 introduces some kinds of MEMS fabrication processes such as the MEMS lithography technology, the bulk silicon technology, the surface silicon technology, the LIGA process and the bonding technology in detail based on many engineering examples. Detailed design idea of the masks is also introduced according to the fabrication process.

In Chapter 5, the micro friction characters of MEMS devices based on micro scale effect are introduced. Based on this theory, the design, simulation, fabrication and experiment for a kind of on-chip micro friction measurement system is mentioned in detail to offer some theoretical help for pertinent research.

Chapter 6 introduces some measuring methods for some kinds of typical parameters including the acceleration measurement, the angular speed measurement, the pressure measurement, the micro-torque measurement and the micro topography measurement.

In Chapter 7, the engineering examples of finite element methods are introduced in detail. This includes the static analysis, the modal analysis, the thermal-structure analysis, the fatigue analysis and the micro capacitor analysis etc. This will give some useful help to pertinent researchers.

This book is written by Associate Professor Guo Zhanshe in Beihang University. Thanks a lot to the subeditor Professor Bai Shaoxian in Zhejiang University of Technology for the writing of Chapter 5 and the modification work, and Ph. D. Huang Manguo in Beijing Changcheng Aeronautical Measurement and Control Technology Research Institute for the writing of Chapter 6. Thanks a lot for the help of my postgraduate students Cheng Fucheng, Li Boyu, Lu Chao, An Ying, Han Jingxuan and Song Ke during the course of translation work. Thanks a lot for the supporting of the Aviation Research Foundation with grant No.: 20153451023.

This book can be used as the reference book for pertinent researchers on MEMS and the teaching book for pertinent postgraduate students.

MEMS is a wide range of research field. Some mistakes and inappropriate content may exist in this book due to the limitation of authors' knowledge. Comments and suggestions from any reader will be appreciated.

Guo Zhanshe
March 2, 2016

Contents

Chapter 1 Introduction 1

  1.1 Concept of MEMS 1

  1.2 Development of MEMS 4

  1.3 MEMS CAD 9

Chapter 2 Basic theory of MEMS 12

  2.1 Theory of electrostatic MEMS comb actuators 12

    2.1.1 Introduction 12

    2.1.2 Operating principles 13

    2.1.3 Plate capacitor theory in ideal condition 14

    2.1.4 The modified model of MEMS plate capaciator 17

    2.1.5 Calculation of electrostatic comb driving force in ideal situation 24

    2.1.6 Weak capacitance detection method of electrostatic comb drive 26

  2.2 Relevant theoretical calculations for the MEMS cantilever beam 34

    2.2.1 Introduction 34

    2.2.2 Theoretical calculation method for cantilever beam 35

    2.2.3 Relevant theoretical calculation of axial tensile and compressive on single-end clamped beams 36

    2.2.4 Related theoretical calculations of double-end clamped beams axial tension and compression 40

  2.3 Membrane theory of MEMS 47

    2.3.1 Theory of clamped around circular diaphragm 48

    2.3.2 Theory of clamped around rectangular flat diaphragm 49

  References 52

Chapter 3 MEMS materials ... 53

3.1 Monocrystalline silicon ... 53
   3.1.1 Introduction ... 53
   3.1.2 Crystal orientation of monocrystalline silicon ... 55
3.2 Polycrystalline silicon ... 64
3.3 Silica ... 66
3.4 Piezoelectric materials ... 67
   3.4.1 Piezoelectric effect and inverse piezoelectric effect of materials ... 67
   3.4.2 Quartz crystal ... 68
   3.4.3 Piezoelectric ceramics ... 73
3.5 Other MEMS materials ... 75
3.6 Summary ... 76

Chapter 4 MEMS technology ... 77

4.1 MEMS lithography process ... 78
4.2 Key technology of MEMS lithography process ... 80
   4.2.1 Wafer cleaning ... 80
   4.2.2 Silicon oxidation ... 80
   4.2.3 Spin coating process ... 87
   4.2.4 Prebaking ... 90
   4.2.5 Exposure ... 92
   4.2.6 Development ... 94
   4.2.7 Hardening ... 96
   4.2.8 Fabrication of the SiO_2 window ... 97
4.3 Subsequent process of MEMS ... 98
   4.3.1 Bulk silicon technology ... 98
   4.3.2 Surface silicon process ... 103
   4.3.3 LIGA technology ... 104
   4.3.4 Sputtering technology ... 105
   4.3.5 Lift-off process ... 107
4.4 Film preparation technology ... 107
4.5 Bonding process ... 108
   4.5.1 Anodic bonding process ... 109

    4.5.2   Silicon-silicon direct bonding ········· 110

    4.5.3   Metal eutectic bonding ········· 113

    4.5.4   Cold pressure welding bonding ········· 114

4.6   Engineering examples of combination for multiple processes to fabricate the MEMS device ········· 115

    4.6.1   Introduction ········· 115

    4.6.2   Engineering example of fabrication process for resonant MEMS gyroscope ········· 115

    4.6.3   Engineering example of electromagnetic micro-motor production process ········· 118

4.7   Summary ········· 124

References ········· 125

Chapter 5   Friction wear and tear under micro scale ········· 126

5.1   Off-chip testing method for micro friction ········· 127

    5.1.1   Micro-tribology test with the pin-on-disc measuring method ········· 127

    5.1.2   Micro-tribology test with AFM ········· 128

    5.1.3   Micro-tribology test with special measuring device ········· 130

5.2   On-chip testing method for micro friction ········· 132

    5.2.1   On-chip testing method actuated by electrostatic force ········· 132

    5.2.2   On-chip micro-friction testing method using the mechanism characters of the bimorph material ········· 139

5.3   Example of the design for an on-chip micro-friction structure ········· 141

    5.3.1   Structure and working principle ········· 141

    5.3.2   Calculation of pertinent theory ········· 142

    5.3.3   Technological analysis of structural design ········· 148

    5.3.4   Testing results and data analysis ········· 152

    5.3.5   Research and test of wear problem of MEMS devices ········· 161

5.4   Summary ········· 165

References ········· 165

Chapter 6   MEMS testing technology and engineering application ········· 169

6.1   Acceleration measurement and corresponding sensors ········· 169

    6.1.1   Working principle of the acceleration sensor and the classification ········· 170

6.1.2　Capacitive silicon micromechanical accelerometer …… 172
6.1.3　Piezoresistive silicon micromechanical accelerometer …… 173
6.1.4　Piezoelectric micromechanical accelerometer …… 174
6.1.5　Resonant silicon MEMS accelerometer …… 175
6.2　Angular speed measurement and corresponding sensors …… 177
6.2.1　Working principle …… 177
6.2.2　Development of MEMS gyroscope …… 178
6.2.3　Classification of micromechanical gyroscope …… 188
6.3　Pressure measurement and corresponding sensors …… 190
6.3.1　Working pincinple …… 190
6.3.2　Resonant silicon micromechanical pressure sensor and its development …… 193
6.4　Measurement of micro-torque …… 198
6.4.1　Introduction …… 198
6.4.2　Working principle of noncontact method …… 199
6.4.3　Theoretical calculation …… 200
6.4.4　Corresponding equipment to realize the noncontact method …… 205
6.4.5　Experiment result and discussion …… 209
6.5　Microscopic morphology testing method …… 211
6.6　Summary …… 212
References …… 212

Chapter 7　Application examples of the finite element method in the design of MEMS devices …… 218

7.1　Important concepts of the software …… 218
7.2　Introduction of the Ansys software interface …… 220
7.3　The coordinate system in Ansys …… 221
7.4　Engineering examples …… 226
7.4.1　Static analysis of single-clamped beam …… 226
7.4.2　Modal analysis of double-clamped beam …… 245
7.4.3　Capacitance analysis of MEMS electrostatic comb fingers drive …… 257
7.4.4　Fatigue strength calculation example …… 264
7.5　Summary …… 272

Chapter 1
Introduction

1.1 Concept of MEMS

MEMS (Micro-ElectroMechanical Systems) is an emerging field developed in recent decades. It's a frontier technology in the twenty-first century based on micro-nanotechnology, and it mainly refers to a technology of design, processing, manufacturing, measurement and control based on silicon materials. It is a micro-system which can integrate mechanical components, optical systems, drive components and electronic control systems into a whole unit. It uses a manufacturing process combined with microelectronics technology and micromachining technology (including silicon bulk micromachining, silicon surface micromachining, LIGA and wafer bonding technologies), to create a variety of excellent performance, low cost, miniaturization of sensors, actuators, drivers and micro-systems. It's a new type of multi-disciplinary technic developed in recent years, which involves machinery, electronics, chemistry, physics, optics, biology, materials and other disciplines.

Currently, the researchers divide the electromechanical system into three categories based on the size of its functional dimensions D: Conventional Electromechanical System, MEMS and NEMS (Nano-ElectroMechanical System). The details are as follows:

Size Range	System Name
$D \geqslant 100~\mu m$	Conventional ElectroMechanical System
$100~\mu m \geqslant D \geqslant 0.1~\mu m$	Micro-ElectroMechanical System
$100~nm \geqslant D \geqslant 0.1~nm$	Nano-ElectroMechanical System

Due to the size effect, many physical properties (including mechanical properties, heat transfer characteristics) of conventional electromechanical systems, MEMS and NEMS have a great difference, and the theoretical research and the process technology are almost completely different. In some MEMS and NEMS, molecular dynamics, quantum mechanics and other microscopic theory have been the primary rationale for the study (shown in Figure 1.1). In NEMS, however, the device processing mainly uses focused ion beam technology, excimer laser direct writing nanometer process technology, and nanoimprint technology, which are rarely used in the traditional electromechanical systems and the microelectromechanical systems.

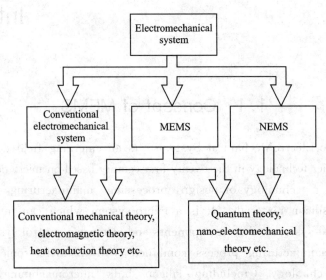

Figure 1.1 The classification of electromechanical systems and the applications of related theories

Compared to the conventional electromechanical system, MEMS has undergone large changes in mechanics and kinematics, material properties, processing technology, detection method and control theory etc. Its main features can be summarized as follows:

(1) Miniaturization: The MEMS device has the advantages of small size and light weight. Usually, the volume of a complete MEMS device is similar to that of a common integrated circuit, and compared with the volume of the sensor and actuator in conventional sense, the volume is usually reduced by more than two orders of magnitude. Thus, in many areas such as the areas with higher requirements for the size and power consumption such as aerospace and other fields, the MEMS device has great advantages.

(2) Integration: MEMS devices can integrate multiple sensors or actuators with different functions and different sensitive directions, forming micro sensor arrays or micro actuator arrays, and they can even constitute more complex microsystems with integrated circuits.

(3) Mass production: With the micro manufacturing process that originates from the semiconductor process, it is able to produce hundreds of thousands of micro electromechanical devices on one substrate at the same time, thus greatly reduces the production costs. Figure 1.2 shows the process layout of a MEMS accelerometer made by bulk silicon technology. As can be seen from this figure, numerous acceleration chips in completely different shapes and sizes are arrayed in a 4-inch-square of the glass, thereby realizing the function that multiple chips can be produced by only one process. Since each sensor chip is produced by once through the same processing technology, the consistency of the unit is good, which can not be achieved by once through traditional machining processing methods.

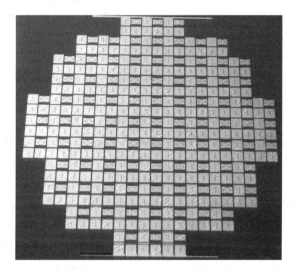

Figure 1.2 The process layout of a MEMS accelerometer

(4) Low cost: The processing technology of MEMS devices is similar to microelectronic technology. Hundreds of thousands of micromechanical parts with good consistency, even the complete MEMS, can be made in a piece of silicon at the same time. Therefore, the manufacturing cost can also be greatly reduced.

(5) Excellent mechanical and electrical properties: The main production material of MEMS devices is silicon, of which the strength, hardness and Young's modulus are similar to those of iron, so the system has good mechanical properties.

(6) Low power consumption: The feature of MEMS devices that have the advantage of small size determines the inevitability of corresponding low power consumption.

The emergence and development of MEMS have improved the miniaturization, versatility, intelligence and reliability level of the information system to a new height. Due to the limits of technic and economic, MEMS was confined to the field of automotive electronics application in the past. But with the volume of MEMS devices becoming smaller and smaller, the price becoming lower and lower and the performance becoming better and better, the design and application space of MEMS devices expands continuously. At present, MEMS has great application prospects in industry, information, communication, national defense, aviation, spaceflight, navigation, medicine, biological engineering, agriculture, environment, family service and other areas. It has attracted the world's attention, as well as a lot of manpower and material resources for development. Therefore, MEMS technology will certainly become another major high-tech industry following the microelectronics technology.

The United States is the earliest country to research and produce MEMS successfully. America's research on MEMS began in the mid-1960s, and attracted widespread attention in the late 1980s. In 1988, American National Foundation invested $1 000 000 to fund MIT, University of California and other 8 universities and Bell Laboratory to begin to carry out the main MEMS research, and the fund increased to $2 000 000 in 1989. Moreover, the fund increased to $5 000 000 in 1993. In 1994, National Key Technology Committee of America listed "micro-, nano-manufacturing" as a national key technology project in the report submitted to the president. Defense Advanced Research Projects Agency of America, which aims at developing dual-use technologies, considered MEMS as a high-tech directly relating to national defense and economic development in 1995 fiscal year plan and focused on its development. Besides, the U. S. Aerospace Company had investigated and evaluated the potential impact of MEMS on future space systems in 1993. Currently American Congress has put MEMS as one of the key disciplines in twenty-first century. Subsequently, Japan, Germany and so on have also carried on the research of MEMS technology. At present, the MEMS technology is involved in almost all fields, and its product sale is increasing at an annual rate of nearly 15 percent.

1.2　Development of MEMS

MEMS appeared along with the development of IC manufacturing technology in 1950s, and developed along with the development of MEMS technology. During this period, the main research contents of MEMS are the physical phenomenon of semiconductor materials and its application in the sensor. In 1954, Smith in Bell Laboratory found the piezoresistive effect of semiconductor silicon and germanium and produced the silicon strain device[1]. In

1959, the famous physicist Feynman gave the landmark speech *There is plenty of room at the bottom* on the famous Physics Annual Conference of America, formally putting forward the imagine of the research of MEMS and the possibility of implementing for the first time, which had a great influence on the development of MEMS. Since the 1960s, MEMS technology got rapid development, and promoted the development of the research technic of MEMS. In the mid-1960s, Bell Laboratory discovered and studied different etching effects of alkali metal solutions produced to different orientations of monocrystalline, and the technology laid the foundation for MEMS wet etching process. Kulite Corporation developed the world's first piezoresistive silicon pressure sensor and acceleration sensor in 1961 and 1970. In 1967, Harvey C. Nathanson put forward the concept of surface passivation technology for the first time in the article *The resonant gate transistor* published in IEEE Transactions on electron devices, and produced the cantilever beam structure with a resonant frequency of 5 000 Hz (shown in Figure 1.3), which marked the emergence of surface passivation technology. In the 1960s, the bonding technology appeared. Nova Sensors Corporation used silicon glass bonding technology to produce pressure sensors, making it possible to produce MEMS chips of multilayered structure. Although these devices had shortcomings such as not being commercial due to the inadequate, these efforts did constitute a part of early outcomes of silicon micromachining technology.

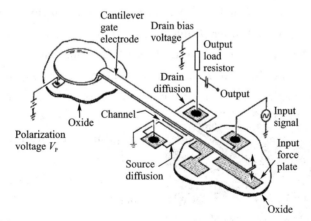

Figure 1.3 The structure of cantilever beam

The 1970s is a period of accelerated development of MEMS. In 1977 and 1979, Stanford University developed the first capacitive pressure sensor and capacitive accelerometer respectively, and started researches of neural probes and silicon chromatographic devices and so on in biomedical field. IBM and HP, using MEMS technology, produced the inkjet printer nozzle respectively in 1977 and 1979[2-3] (shown in Figure 1.4). At present, the printer nozzle is still one of the most important products in MEMS field, marking the beginning of the application of MEMS technology.

Honeywell, Motorola and some other companies also introduced a mass production of pressure and acceleration sensors in the late 1970s. The 1980s is also a period of rapid development of MEMS, in which countries around the world began to study in the field of MEMS, therefore manufacturing technology

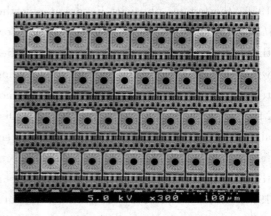

Figure 1.4　Inkjet printer nozzle

continued to emerge and get perfect, applications continued to expand, researches on basic theory and design methodology continued to go further. In 1985, Howe and other people in UCB created the polycrystalline silicon resonant beam integrated with the MOS circuits, proving the compatibility between polysilicon and IC technology. Subsequently, UCB, MIT and University of Wisconsin perfected the sacrificial layers micro processing technology, and produced the complex MEMS system successfully, which laid on the foundation of MEMS and IC integration, and was named "surface micromachining" technology by Barth in HP in 1985. In 1987, Fan L. S. and other people in University of California, Berkeley created the electrostatic micro-motor[4] (shown in Figure 1.5) for the first time with the surface micromachining technology. The motor was 120 μm in diameter, 1 μm in thickness, and its speed can reach 500 rpm, which was the first time to realize Feynman's plan. Though micro-motor hasn't got the application so far, it played a great role in promoting the large-scale emergence of MEMS around the world then, and promoted the MEMS research to a new height. In 1984, University of Michigan invented the dissolved silicon technology based on silicon glass bonding and thick boron block KOH etching, and the resonant silicon micromechanical pressure sensor came out in the same year. In 1979, Stanford University in America used

Figure 1.5　The structure chart of electrostatic micro-motor

micromachining technology and silicon materials to create the open loop direct frequency output accelerometer for the first time; In 1989, Satchell D. W. and Greenwood J. C. used methods of thermal excitation and piezoelectric detection to design the three-beam structure silicon directly frequency output accelerometer[5] for the first time; In 1990, Chang C. S. and other people[6] used the double beam-structure to design a direct frequency output accelerometer which is sensitive in both directions, and they applied for a patent; LIGA (Lithographie, Galvanoformung and Abformtechnik) processing technology was born in Germany in 1985. The process can not only make high aspect ratio three-dimensional structure (shown in Figure 1.6), but also make it possible to manufacture complex three-dimensional metal structure. In 1986, Binning in IBM invented the atomic force microscopy based on micromachining technology, achieving the measurement of light force at Newton, even the nano-Newton power level and the nano surface, and the research group then won the Nobel Prize. On the conference Micro-Tele-Operated Robotics held in Salt Lake City in 1989, Howe in UCB suggested using MEMS as the name of this field, which was the first time the name of MEMS was determined.

Figure 1.6 Micro gear structure made by LIGA technology

In the 1990s, MEMS technology entered a period of high input and output. Countries around the world spent a lot of money on MEMS research and made a great progress, and many MEMS products came out. At the same time, the level of processing also reached a very high level. The Summit V surface micromachining technology developed by American National Laboratory Sandia could manufacture polycrystalline silicon micromechanical structure of 5 layers, and had achieved complex resonators, motors, gears, adjustable micromirrors and other devices, which represented the highest level of micromachining in the world. Figure 1.7 shows the scanning electron microscopy image of the adjustable micromirror of

MEMS made for the mechanism, and the structure is still the highest level of the process at present.

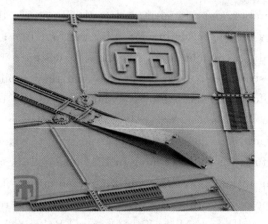

Figure 1.7 The scanning electron microscopy image of the micromirror of MEMS

Draper Laboratory introduced the double-frame bulk silicon micromechanical vibrating gyroscope sample driven by plate capacitor in 1988. It was the first silicon micromechanical gyroscope, and the plane dimension was 300 μm $\times$ 600 μm[7]. In 1993, ADI Corporation in America achieved the commercialization of MEMS acceleration sensors. In the next 10 years, the company introduced a series of acceleration sensors, which have been widely used in the field of automotive electronics and aerospace. In 1988, Greenwood J C published his research findings[8], making substantial content and theoretical analysis of resonant silicon micromechanical pressure sensor designed by himself in 1984. The design was then used by Druck Corporation and achieved industrialization by marketing in 1990s. As shown in Figure 1.8, it used a butterfly structure. When applying tested pressure to the back of the silicon diaphragm, the diaphragm is deformed, resulting in strain in the wedge part of butterfly resonant beam. Thus the natural frequency of the resonant beam changes, and the pressure can be calculated by testing its change value. The sensor, using the bulk silicon technology, could be used to measure the flying height of the aircraft and was widely used.

In 2002, University of California, Berkeley invented the resonant silicon micromechanical gyroscope[9-10], making the performance of the gyro further improved. At present, MEMS has formed the industrialization, and its products include acceleration sensors, gyroscopes, inkjet print heads, pressure sensors and so on, and its application fields refer to aerospace, automotive electronics, the game industry, mobile phones and so on.

Figure 1.8 The scanning electron microscopy image of the butterfly resonator

1.3 MEMS CAD

MEMS technology is a large multi-field discipline, and it has its own inherent theory. The corresponding MEMS device is a new kind of device, whose design contains the exploration and development of new work and new structures, rather than mature traditional designs. Currently, there is a great difficulty in the design of MEMS devices, especially some of the resistance on the object at a microscopic scale is still in a state of uncertainty, and a lot of transformation produced by load must be analyzed by nonlinear theory. It makes people must rely on computers to solve the design problems of MEMS. Currently, with the two or three dimensional computer graphics technology and the finite element analysis technique, we can carry out the design and simulation of complex MEMS structures and graphics, realizing the simulation and design optimization of the structure, properties of the MEMS device. Moreover, with the rapid development of MEMS, MEMS CAD (Computer Aided Design of MEMS) has become the essential technology in the research, and the related research has become a hotspot as well.

In the early 1990s, computer aided design software MEMS CAD which was used in the design of silicon pressure sensors was introduced in many countries such as America. In 1996, Microcosm Corporation introduced the first set of commercialized professional design analysis software of MEMS devices of the world—MEMS CAD. In 2001, Microcosm Corporation was renamed Coventor Corporation, and the software was renamed Coventor Ware correspondingly, which is the world's most powerful and largest professional software of MEMS currently. The software has dozens of specialty modules, and it's convenient to realize the architectonics, technology, simulation and so on of MEMS devices. The software has advantages of powerful modules, abundant database and process data, easy operation, perfect data interface with other software and so on. In the light of the characteristics of MEMS, this book focuses on another important MEMS CAD software—Ansys, introduces its applications in MEMS design and gives the engineering examples correspondingly, makes every effort to validate the feasibility of the design in advance, reduces problems existing in the design process, and achieves the goal of shortening the design time, reducing production costs and improving production efficiency. Ansys is a large and commonly used finite element software which integrates analysis of structure, fluid, electric field, magnetic field and sound into a whole (Figure 1.9 shows its functions). The software can not only do simulation calculation conveniently, but also make the interface with most of CAD software and carry out the data sharing and exchange, for example, Pro/Engineering, NASTRAN, Alogor, i-deas, Solidwork, etc. The software is one of the advanced CAD tools of modern product design, and it mainly includes three parts: pre-processing module, analysis and calculation module and post-processing module.

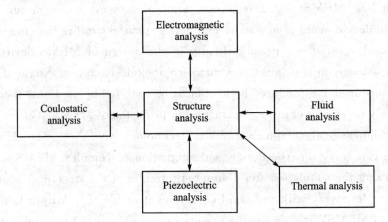

Figure 1.9 Ansys multi-physics coupling figure

Pre-processing module provides a powerful entity modeling and meshing tool, through which users can easily construct the finite element model. This part mainly includes 4 steps: unit type classification, material parameters setting, structural entity modeling and meshing.

Analysis and calculation module includes structure analysis (linear analysis, nonlinear analysis and highly nonlinear analysis can be performed), fluid dynamics analysis, electromagnetic analysis, sound field analysis, piezoelectric analysis and the physical field coupling analysis, and it can simulate the interaction of a variety of physical media with the ability of sensitivity analysis and optimization analysis. The module mainly includes the definition analysis type, the analysis option, the loading data and the loading step option, and then begins the finite element solution.

Post-processing module includes two parts: general post-processing module POST1 and time history post-processing module POST26. Post-processing module can display calculation results in color contour, gradient, vector, particles tracing, three-dimensional slice, transparent and semi-transparent (the interior structure can be seen) and other graphics modes, and it can also display or output calculation results in the form of charts and curves. The results may include displacement, temperature, stress, strain, speed and heat flux and so on, and the output modes include graphical display and data list. Figure 1.10 shows the operation interface of Ansys.

Based on its powerful functions, Ansys has been widely applied in computer aided design of MEMS. Many MEMS devices, such as MEMS acceleration sensors, MEMS gyroscopes, comb driver analysis (shown in Figure 1.11), radio frequency devices, analysis of resonator, biochips, micro lens technology and so on, have made use of the software to verify the feasibility in the design process, which will shorten the design time and reduce the cost of products greatly.

Figure 1.10 The graphical interface of Ansys

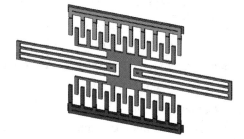

Figure 1.11 The comb drive model built by Ansys

Chapter 2
Basic theory of MEMS

This chapter takes the most commonly used components in MEMS as study objects, and introduces them based on related theories to provide theoretical support for related designs, including theories of electrostatic comb actuators, cantilevers and diaphragms.

2.1 Theory of electrostatic MEMS comb actuators

2.1.1 Introduction

As a typical MEMS component, the electrostatic MEMS comb actuator has two applications in the micro-electromechanical system. One is as the significant power source to drive the system; the other application is as the important stimulate-detection component in many systems such as sensor. Although the driving force of electrostatic actuation is low, it uses voltage control, and has low energy dissipation, short response time, high efficiency and good technology compatibility; it can be processed by bulk silicon and surface micro-mechanical technology, and easy to realize system integration. It is the most commonly used component in MEMS, which has the following advantages:

(1) Normally it is used for variable area detection and overcomes the shortcoming of variable gap detection that can only detect infinitesimal displacement. It can either detect extensiveness or subdivide within pole pitch to reach enough precision and resolution ratio.

(2) It can greatly increase capacitance with the increasing of comb teeth numbers.

(3) It is easy to form elaborate geometric constructions such as differential detection, and is very beneficial to the improvement of detection precision.

(4) It has low calorific value and weak interference to sensor.

(5) Detection port has simple construction that is easy to integrate with semiconductor

technology, and beneficial to device manufacture.

(6) It is non-contact detection, and will not detect the disturbing force to inertial mass which has infinitesimal displacement, in order to keep the measuring result precise.

(7) It uses non-contact capacitor drive-detect pattern, and has low temperature affection, good test stability and high precision, thus provides simple and excellent way to realize the closed-loop control of micro inertial device.

At present, MEMS electrostatic comb actuator has been widely used in many MEMS sensors. Almost all MEMS acceleration sensors, spinning tops and pressure sensors use the electrostatic comb actuator for their sensitive elements' stimulate and detect methods (shown in Figure 2.1).

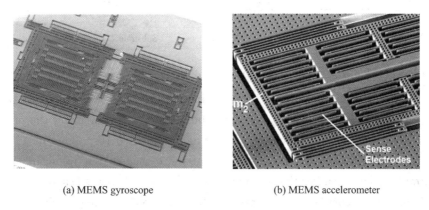

(a) MEMS gyroscope   (b) MEMS accelerometer

Figure 2.1　Electrostatic comb drive actuators in MEMS inertial device

2.1.2　Operating principles

The basic working principle of electrostatic comb construction is shown in Figure 2.2. It consists of movable comb teeth, fixed comb teeth, supporting beam, anchor and the substrate.

Fixed comb teeth unify the fixed electrode pad, while movable comb teeth unify the movable electrode pad which connects with the fixed island through supporting beams and fixes on substrate, then movable comb teeth flat suspends in midair through the fixed island and supported beam in the middle to decrease frictional resistance of the system.

If a driving voltage V_d is applied between the driving end and movable comb teeth, the electrostatic force will be initiated between the comb teeth which are in proportion to the capacitance of the comb teeth. When detecting element is in measuring conditions, the

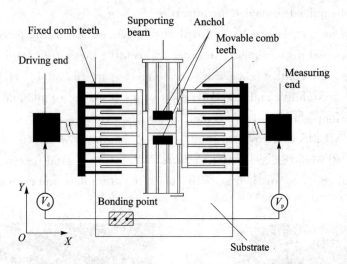

Figure 2.2 The basic working principle of electrostatic comb actuator

supporting beam will initiate deformation and the overlap area of comb teeth will change. The value of the capacitance will change correspondingly. By measuring the change of the capacitance, the parameters to be measured will be obtained correspondingly.

It is concluded from the working principle of the device that if we want to realize the drive or stimulate-detection of the system, to study the theory of plate capacitor is an essential. Thus, the theory of electrostatic comb teeth around plate capacitor will be carried out. As for the special character of MEMS fabrication process, traditional theoretical calculation has obvious differences, it is necessary to effectively rebuild the theoretical model.

2.1.3　Plate capacitor theory in ideal condition

2.1.3.1　Ideal capacitance theory with no relative displacement

Theory of ideal comb teeth capacitance is mostly used in MEMS for preliminary performance estimation. Because the edge effect is ignored and two plates are strictly parallel (shown in Figure 2.3), we can calculate the capacitance according to

$$C = \frac{\varepsilon s}{\delta} = \frac{\varepsilon_0 \varepsilon_r ab}{\delta}. \tag{2.1}$$

In this equation:

ε_0 is the dielectric constant in vacuum(F/m), $\varepsilon_0 = \dfrac{10^{-9}}{4\pi \times 9}$ F/m;

ε_r is the relative dielectric constant between counter electrodes, $\varepsilon_r = \dfrac{\varepsilon}{\varepsilon_0}$, which is 1 in air;

a is the width of comb teeth;

b is the overlapping length of adjacent comb teeth;

δ is the gap between adjacent comb teeth.

In practical use, there is a large error between detection result and ideal computing result in capacitance of electrostatic comb construction.

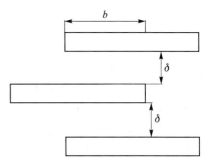

Figure 2.3 Model of plate capacitor theory

It is important to study key factors which will influence the calculation accuracy of electrostatic comb. Then, deduce high-precision correction calculation model which has overall consideration of all kinds of influence factors.

Considering of configuration form of MEMS plate capacitors, we should make relative movement if we want to realize its use of detection or drive in MEMS devices. According to direction of motion, there are two forms—interval variation and area variation.

2.1.3.2 The sensitive element of variable interval capacitor's theory

It is concluded from equation (2.1) that C is inversely proportional to the gap between counter electrodes, and has big non-linearity. So when they are working, the movable electrodes can only work in a small range.

When the gap δ decreases by $\Delta\delta$ and changes to $\delta - \Delta\delta$, the capacitance C will add ΔC:

$$\Delta C = \frac{\varepsilon S}{\delta - \Delta\delta} - \frac{\varepsilon S}{\delta} \qquad (2.2)$$

And

$$\frac{\Delta C}{C} = \frac{\dfrac{\Delta\delta}{\delta}}{1 - \dfrac{\Delta\delta}{\delta}} \qquad (2.3)$$

When $\dfrac{\Delta\delta}{\delta} \ll 1$, we can change equation(2.3) into series type as follow:

$$\frac{\Delta C}{C} = \frac{\Delta\delta}{\delta}\left[1 + \frac{\Delta\delta}{\delta} + \left(\frac{\Delta\delta}{\delta}\right)^2 + \cdots\right] \qquad (2.4)$$

It is concluded from equation (2.4) that the relative change of capacitance can be expressed as multinomial of the gap's relative change. Orders of the result depend on the selection of the precision. If we want to keep the linearity between them, their relationship can be expressed as

$$\frac{\Delta C}{C} \approx \frac{\Delta \delta}{\delta}\left(1 + \frac{\Delta \delta}{\delta}\right) \tag{2.5}$$

For sensitive element of capacitance with variable gaps, equation (2.5) can be deduced as the expected linear relation. Equation (2.5) can be deduced with non-linear curve.

From the analysis above, we know:

(1) If we want to improve sensitivity K, we should decrease initial interval δ, but breakdown voltage of capacitor and the difficulty of assembly work should be taken into consideration.

(2) Non-linearity will rise with the rise of relative displacement, so we should limit the relative displacement of movable electrodes to ensure linearity. Usually set $|\Delta \delta_m / \delta|$ as $0.1 \sim 0.2$.

(3) To improve non-linearity, we can use the differential type as Figure 2.3 shows. When one capacitance increases, its characteristic equation is equation (2.4); the other capacitance will decrease. We can acquire a very good character if we change circuit type properly. We can use bridge circuit as processing circuit to improve precision.

The sensitive element of variable gap capacitor is widely used in MEMS such as acceleration sensor with variable gaps.

2.1.3.3 The sensitive element of variable area capacitor's theory

Figure 2.4 shows the schematic diagram of the sensitive element which is parallel plate capacitor of movable area. When regardless of edge effect, its characteristic equations of capacitance are

$$C = \frac{\varepsilon b(a - \Delta x)}{\delta} = C_0 - \frac{\varepsilon b \Delta x}{\delta} \tag{2.6}$$

$$K = \frac{\Delta C}{\Delta x} = \frac{\varepsilon b}{\delta} \tag{2.7}$$

Capacitive sensor of movable area is linearity, sensitivity K is a constant, and when b increases or δ decreases, K increases. The width of electrodes will not influence K, but will influence edge effect.

Compared with the sensitive element of variable interval capacitor, the sensitive element of variable area capacitor has advantages as follows:

(1) The relative change of capacitance and relative displacement is linearly, so sensors use this method to have high output signal detection precision, and handle the signal with ease.

(2) The air damping of two electrodes is slide-film damping, which will decrease the coefficient of air damping. This can improve the quality factor greatly for vibrating sensor.

(3) In design process the electrodes can produce quite big linear relative displacement, and lead to big change of linear capacitor signal. Enhance of output signal can largely reduce the difficulty in weak signal detection of sensors.

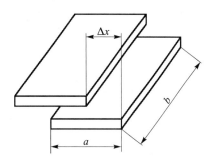

Figure 2.4 Structure chart of movable area comb teeth

Since the capacitive sensor of movable area has so many advantages, it has become widely used in MEMS.

2.1.4 The modified model of MEMS plate capacitor [1-2]

In ideal situation, all comb teeth in electrostatic comb teeth acturtor are paralleled. However, because of the influence of the fabrication process, the comb teeth are unparalleled in reality. On the other hand, the depth-to-width ratio of electrostatic comb actuator structure is much less than original plate capacitor, so its edge effect cannot be ignored. These will affect the computing of electrostatic comb teeth's capacitance, so it must be studied carefully.

2.1.4.1 Taking fabrication process error into consideration

Restricted by the fabrication process of MEMS device, there is always some errors between actually made electrostatic comb teeth and ideal model. The uneven of comb teeth surface profile mainly affects its mechanical properties; and the shortage of perpendicularity, or the small included angle between the comb teeth and the vertical plane, mainly affects its electrical properties. It is necessary to study the effect of the small included angle between the comb teeth and the vertical plane on precision of capacitance computing. In this situation, a sectional view of a pair of adjacent comb teeth is shown in Figure 2.5.

As the space between adjacent comb teeth is a stably changed variable, we consider to use infinitesimal method in calculus theory to calculate the capacitance. We can use multiple method or series method according to the differential of selected infinitesimal.

1. Multiple method

Suppose the extension cord of a pair of comb teeth intersected at the cross point, O and

17

the intersection angle is 2θ. Divide it into infinite small elements in crosswise, and select a pair of small elements symmetric as shown in Figure 2.6.

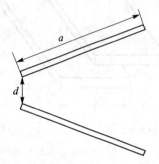

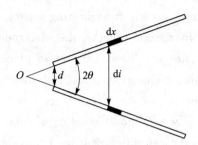

Figure 2.5 Sectional view of a pair of adjacent comb teeth

Figure 2.6 Use multiple processes to select small element

Because the length of selected elements is infinitesimal, we can consider their intervals are fixed. Regarding the capacitor consisting of a pair of small elements as ideal parallel-plate capacitor and supposing its length is dx, interval is di, its capacitance is

$$c_i = \frac{\varepsilon b \cdot dx}{di} = \frac{\varepsilon b \cdot dx}{2x \cdot \tan\theta} \qquad (2.8)$$

The capacitor consisting of a pair of adjacent comb teeth is composed by all these small elements of capacitors in parallel, so its capacitance is

$$c_u = c_1 + c_2 + \cdots + c_i + \cdots \qquad (2.9)$$

If equation (2.9) is integrated in the range of $\left[\frac{d}{2\tan\theta}, \frac{d}{2\tan\theta} + a\right]$, we can calculate the capacitance of a pair of adjacent comb teeth as

$$c_u = \int_{\frac{d}{2\tan\theta}}^{\frac{d}{2\tan\theta}+a} \frac{\varepsilon b \cdot dx}{2x \cdot \tan\theta} = \frac{\varepsilon b}{2\tan\theta} \cdot \ln\left(1 + \frac{a \cdot 2\tan\theta}{d}\right) \qquad (2.10)$$

2. Series method

Suppose the extension cord of a pair of comb teeth is intersected at the cross point O, and the angle between them is 2θ. Dividing it into infinite small elements in lengthways, the calculating model is shown in Figure 2.7.

Because the interval of selected pair of small elements is infinitesimal, we can consider their intervals are fixed. Regarding a pair of small elements are ideal capacitor and their gap is di, its capacitance can be expressed as

$$c_i = \frac{\varepsilon ab}{di} = \frac{\varepsilon ab}{r \cdot d\varphi} \qquad (2.11)$$

Because the capacitor is composed of all of the series connected little elements, the total

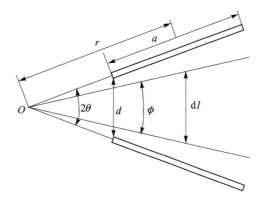

Figure 2.7 Using series processes to select the small element

capacitance can be expressed as

$$\frac{1}{c_u} = \frac{1}{c_1} + \frac{1}{c_2} + \cdots + \frac{1}{c_i} \qquad (2.12)$$

Using the integrating method in the range of $[0, 2\theta]$, reciprocal of the calculated capacitance can be expressed as

$$\frac{1}{c_u} = \int_0^{2\theta} \frac{1}{c_i} = \int_0^{2\theta} \frac{r \cdot d\varphi}{\varepsilon a b} = \frac{2r\theta}{\varepsilon a b} \qquad (2.13)$$

From the geometrical relationship shown in Figure 2.7, we know that $r = \frac{a}{2} + \frac{d}{2\sin\theta}$, so

$$c_u = \frac{\varepsilon a b}{\left(a + \dfrac{d}{\sin\theta}\right)\theta} \qquad (2.14)$$

In fact, the small element divided in multiple processes can't be considered as ideal parallel-plate capacitor from physics definition. Because the length of electrode is much less than the gap, the theory is in contradiction with the definition of ideal parallel-plate capacitor. So the result calculated by series method is confirmed to physics definition.

When θ is close to 0, we know from equation (2.14) that

$$c_u = \lim_{\theta \to 0} \frac{\varepsilon a b}{\left(a + \dfrac{d}{\sin\theta}\right)\theta} = \frac{\varepsilon a b}{d} \qquad (2.15)$$

It is concluded from equation (2.15) that when the comb teeth are in parallel, it fits the ideal condition, and correction calculation model equation (2.14) is still effective. It does not contradict the ideal calculation model equation (2.10).

The relative error between the ideal calculated result and the corrected result is

$$\delta_u = \frac{|c_{\text{ideal}} - c_u|}{c_{\text{ideal}}} \times 100\% = \left[1 - \frac{d}{\left(a + \dfrac{d}{\sin\theta}\right)}\right] \times 100\% \tag{2.16}$$

Taking the lab fabricated MEMS comb teeth actuator as the specimen (micrograph is shown in Figure 2.8 and the detailed dimensions are shown in Table 2.1), the calculated result is shown in Figure 2.9. During the course of the calculation, the change of the angle is supposed to be in the range of 0° to 2° (the perpendicularity of comb teeth varies between 88° and 90°).

Table 2.1 Size parameters of electrostatic comb teeth drive

Length/μm	Width/μm	Overlapping length/μm	Interval/μm	Thickness/μm	Included angel/(°)	Pairs
100	72	50	4.5	4	0.5	90

Figure 2.8 Sketch of the MEMS comb teeth actuator

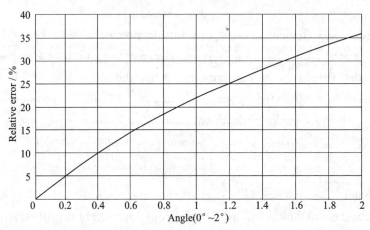

Figure 2.9 Relative error between ideal value and correction value considering craftsmanship error

It is concluded from Figure 2.9 that when the included angel θ between comb teeth and vertical plane is increasing slightly, relative error δ_u between ideal value and correction value considered fabrication process error will increase quickly. When the value of included angel θ is only 0.4°, relative error reaches 10%. Thus we know the fabrication process that causes unevenness has great effect on the calculation of the capacitance. It can't be ignored.

2.1.4.2 Theoretical model considering the edge effect

According to physical definition, the ideal parallel-plate capacitor means a capacitor consists of two paralleled arranged infinite plate conductors, which means the gap of the two electrodes is constant, and the length and width of the electrode are much bigger than the gap. In this situation, the edge effect of electrode can be ignored, which means power line between electrodes is equally distributed.

However, for actually fabricated electrostatic comb teeth, the values of length and width of comb teeth are limited, so the edge effect in two directions can't be ignored. The power lines between adjacent comb teeth can not be limited in the area between electrodes. They are straight lines only in the middle area. The power line which is not in the middle area is outward, and it is more outward if the power line is nearer to the edge of electrode. At the edge of comb teeth, power line's curve degree is the most serious. The power lines between a pair of comb teeth are shown in Figure 2.10.

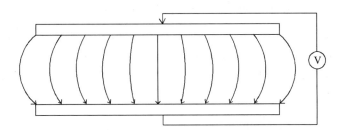

Figure 2.10 Schematic diagram of power lines between a pair of comb teeth

We firstly consider edge effect in length direction. According to conformal mapping theory, for a capacitor consists of a pair of adjacent comb teeth, the capacitance of per unit length is

$$c_j = \frac{\varepsilon a}{d} + \frac{\varepsilon}{\pi}\left\{1 + \ln\left[1 + \frac{\pi a}{d} + \ln\left(1 + \frac{\pi a}{d}\right)\right]\right\} \qquad (2.17)$$

When only considering this direction's edge effect, the capacitance of the capacitor consisting of a pair of adjacent comb teeth is

$$c_{l1} = c_j \cdot b = \frac{\varepsilon ab}{d} + \frac{\varepsilon b}{\pi}\left\{1 + \ln\left[1 + \frac{\pi a}{d} + \ln\left(1 + \frac{\pi a}{d}\right)\right]\right\} \qquad (2.18)$$

According to equations (2.17) and (2.18), the relativity error between ideal calculating result and correction calculating result considering edge effect in length direction is

$$\delta_{l1} = \frac{|c_{\text{ideal}} - c_{l1}|}{c_{\text{ideal}}} \times 100\% = \frac{d}{\pi a}\left\{1 + \ln\left[1 + \frac{\pi a}{d} + \ln\left(1 + \frac{\pi a}{d}\right)\right]\right\} \times 100\% \qquad (2.19)$$

where $\frac{a}{d}$ is the depth-to-width ratio of electrostatic comb teeth structure.

When $\frac{a}{d}$ approaches ∞, we have

$$\delta_{l1} = \lim_{\frac{a}{d} \to \infty} \frac{d}{\pi a}\left\{1 + \ln\left[1 + \frac{\pi a}{d} + \ln\left(1 + \frac{\pi a}{d}\right)\right]\right\} \times 100\% = 0 \qquad (2.20)$$

From equation (2.20) we know that when depth-to-width ratio approaches to infinitely big, the parallel arranged comb teeth are considered to be the ideal plate capacitor. The relative error between correction model and ideal computing model is zero, which agrees with ideal calculating model.

The aspect ratio is an important parameter in MEMS fabrication process. Therefore, we can study the impact of edge effect in the calculation accuracy of electrostatic comb structure's capacitance based on it.

According to the current fabrication conditions for bulk silicon technology, $\frac{a}{d}$ is supposed to be in the range of 5 to 25. According to equation (2.19), the relative error δ_{l1} between ideal calculation results and correction calculation results considering the edge effect of comb in length direction can be caculated out by using MATLAB. The curve of it is shown in Figure 2.11.

Figure 2.11 shows that when aspect ratio $\frac{a}{d}$ increases, relative error δ_{l1} of ideal calculated results and corrected results considering the edge effect of comb in length direction decreases, when aspect ratio $\frac{a}{d}$ reaches 20 (almost the limit of silicon MEMS fabrication process), relative error δ_{l1} is still greater than 8%. And in practice, the edge effect in the width direction of the comb teeth should also be taken into consideration. From it we know that the edge effect has significant influence on capacitance calculation precision of electrostatic comb structure and can't be ignored.

Similarly, when only the edge effect in the width direction of the comb teeth is taken

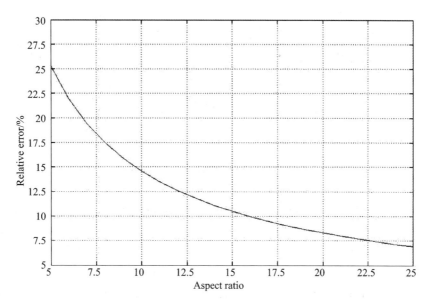

Figure 2.11 Relative error between ideal calculation results and correction calculation results considering the edge effect

into consideration, the capacitance of a capacitor constituted by a pair of adjacent comb can be expressed as

$$c_{l2} = \frac{\varepsilon ab}{d} + \frac{\varepsilon a}{\pi}\left\{1 + \ln\left[1 + \frac{\pi b}{d} + \ln\left(1 + \frac{\pi b}{d}\right)\right]\right\} \quad (2.21)$$

Combining equations (2.18) and (2.21), we can know that when considering edge effect in two directions, the capacitance of a capacitor constituted by a pair of adjacent comb can be expressed as

$$c_l = \frac{\varepsilon ab}{d} + \frac{\varepsilon a}{\pi}\left\{1 + \ln\left[1 + \frac{\pi b}{d} + \ln\left(1 + \frac{\pi b}{d}\right)\right]\right\} + \frac{\varepsilon b}{\pi}\left\{1 + \ln\left[1 + \frac{\pi a}{d} + \ln\left(1 + \frac{\pi a}{d}\right)\right]\right\} \quad (2.22)$$

2.1.4.3 The establishment of the model considering fabrication process and edge effect

For electrostatic comb teeth actuators in practical application, its capacitance is effected by fabrication process and edge effect at the same time. Errors between actual test results and ideal calculation results are the result of these two aspects' influences. Therefore, to achieve high-precision calculation, we have to consider influences in two aspects.

Combining equations (2.19) and (2.22), derive correction calculation model of electrostatic comb structure considering manufacturing process error and edge effect as follow:

$$C_{\text{total}} = \frac{\varepsilon ab}{\left(a+\dfrac{d}{\sin\theta}\right)\theta} + \frac{\varepsilon a}{\pi}\left\{1+\ln\left[1+\frac{\pi b}{\left(a+\dfrac{d}{\sin\theta}\right)\theta}+\ln\left(1+\frac{\pi b}{\left(a+\dfrac{d}{\sin\theta}\right)\theta}\right)\right]\right\} +$$

$$\frac{\varepsilon b}{\pi}\left\{1+\ln\left[1+\frac{\pi a}{\left(a+\dfrac{d}{\sin\theta}\right)\theta}+\ln\left(1+\frac{\pi a}{\left(a+\dfrac{d}{\sin\theta}\right)\theta}\right)\right]\right\} \qquad (2.23)$$

2.1.5 Calculation of electrostatic comb driving force in ideal situation

Electrostatic comb drives' electrostatic force can be divided into two parts: normal electrostatic force (shown in Figure 2.12) and tangential electrostatic force (shown in Figure 2.13) according to its direction of action. The direction of tangential electrostatic force can be divided into two kinds according to its tangential direction: overlapping direction along the comb and vertical direction with it. They can be respectively expressed as vectors **L** and **W**, the forces corresponding to them are F_L and F_W.

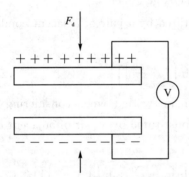

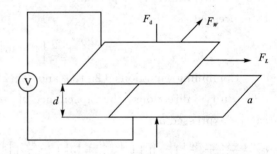

Figure 2.12 Calculation of electrostatic force in electrostatic comb drive's normal direction

Figure 2.13 Tangential electrostatic force of electrostatic comb drive's calculation

2.1.5.1 Calculation of drive's displacement under electrostatic force

As the main driving source of MEMS systems, during the design process we will calculate electrostatic comb drives driving force (confer the structure shown in Figure 2.2), to achieve the evaluation of overall system's performance. From Figure 2.2 we know that an electrostatic comb drive system generally has a supporting beam for support, the supporting mechanism can be equivalent to a spring system. Assuming its stiffness is k. When a driving force is applied to the comb drive, its balance equation can be expressed as

$$F = kx \tag{2.24}$$

where F represents the driving force of electrostatic comb drive, x represents the displacement under the action of electrostatic force.

2.1.5.2 Calculation of the electrostatic driving force perpendicular to the direction of the capacitor plates

Calculation of the driving force between the plates is carried out by using the energy method. From equation (2.1) we know that when voltage V is applied to a paralleled-plate capacitor, its capacitance can be expressed as

$$c = \frac{\varepsilon n a b}{\delta} \tag{2.25}$$

According to energy method, when certain potential V is applied to two paralleled plates, corresponding electric potential energy can be expressed as

$$U = \frac{1}{2}cV^2 = \frac{\varepsilon n a b V^2}{2\delta} \tag{2.26}$$

According to derivative theories, the electrostatic force perpendicularly to the plates can be calculated by derivative of the potential to displacement in the vertical direction as

$$F_d = \frac{\partial U}{\partial \delta} = \frac{1}{2}\frac{\varepsilon n a b V^2}{\delta^2} \tag{2.27}$$

2.1.5.3 Calculation of electrostatic driving force parallel to the direction of the capacitor plates

Electrostatic driving force parallel to the direction of the capacitor plates can be divided into the comb overlapping direction and the direction perpendicular with it, which can be respectively expressed as vectors **L** and **W**, and the forces corresponding to them are F_L and F_W (shown in Figure 2.13). Similarly, using energy derivative to displacement method, we can get the following expressions of driving forces in two directions above:

Width direction $\qquad F_W = \frac{\partial U}{\partial a} = \frac{1}{2}\frac{\varepsilon n b V^2}{\delta} \tag{2.28}$

Length direction $\qquad F_L = \frac{\partial U}{\partial a} = \frac{1}{2}\frac{\varepsilon n a V^2}{\delta} \tag{2.29}$

As can be seen from the above two equations, F_W has nothing to do with the width a, meanwhile F_L has nothing to do with the length b.

Similarly, during the design process, since the driving force of capacitance device with variable area and displacement is in a linear relationship, it has become the predominant mode of capacitive drive design.

2.1.5.4 Calculation of electrostatic driving force under the action of a sinusoidal alternating voltage

In practical engineering design, in order to ensure an electrostatic comb drive achieving reciprocating motion, the driving voltage applied to it typically is a driving alternating voltage that has a certain DC bias, thus suppose the voltage is

$$V = V_0 + V_1 \sin \omega t \tag{2.30}$$

Taking equation (2.30) into equation (2.28), we get

$$F_W = \frac{\partial U}{\partial a} = \frac{1}{2} \frac{\varepsilon n b V^2}{\delta} = \frac{1}{2} \frac{\varepsilon b n (V_0 + V_1 \sin \omega t)^2}{\delta} \tag{2.31}$$

Expand and sort out

$$F_W = \frac{n \varepsilon b}{2d} (V_0^2 + 2V_0 V_1 \sin \omega t + V_1^2 \sin^2 \omega t) \tag{2.32}$$

In the process of applying voltage, set $V_1 \ll V_0$, and in equation (2.32), $V_1^2 \sin^2 \omega t$ can be ignored, so simplify equation (2.32) as

$$F_W \approx \frac{n \varepsilon b}{2d} (V_0^2 + 2V_0 V_1 \sin \omega t) = \frac{n \varepsilon b V_0^2}{2d} + \frac{n \varepsilon b V_0 V_1 \sin \omega t}{d} \tag{2.33}$$

As can be seen in equation (2.33), under the action of alternating voltage, the electrostatic force applied to the comb drive can be equivalent to a electrostatic force which has a alternating bias. So, combining equation (2.33) and (2.24), we can calculate the displacement of the comb drive's expression in alternating force as

$$x = \frac{F_W}{k} = \frac{1}{k} \left(\frac{n \varepsilon b V_0^2}{2d} + \frac{n \varepsilon b V_0 V_1 \sin \omega t}{d} \right) \tag{2.34}$$

Similarly, we can calculate the drive's displacement in other two directions.

2.1.6 Weak capacitance detection method of electrostatic comb drive

Weak capacitance detection technology of electrostatic comb drive is one of the key technologies of MEMS comb drive. Currently, there are several common weak capacitive sensing technologies, and the main goal is to convert capacitance into frequency, current or voltage output by using various signal conditioning circuit. As a result, they are classified according to the principles of conversion, such as capacitance-frequency method, switched capacitor method, AC bridge method, ring diode method and modulation and demodulation method.

2.1.6.1 Capacitance-frequency method

Capacitance-frequency method changes the capacitance into frequency output. Variation of output signal's frequency represents the variation of capacitance to be measured. This method can use low power consumption oscillating circuit[4] or the charge-discharge of capacitance to be measured to control the on-off of trigger. It also can be realized by using operational amplifier circuit.

Figure 2.14 shows the multi-vibrator with timer NE555 as key component, and capacitor C_x to be measured is in the circuit as one of the timing components.

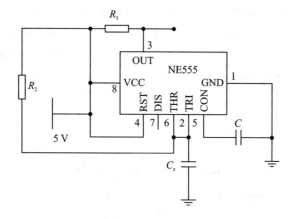

Figure 2.14 Capacitance-frequency convert circuit with timer NE555

The output square wave signal of the circuit's frequency is

$$f = \frac{1}{2R_1 C_x \ln 2} \qquad (2.35)$$

The advantages of capacitance-frequency method are obvious. They can be summarized as follows:

(1) The theory is simple. It can be realized by small components.

(2) The output is frequency signal. It can be converted to digital signal easily, which is easy to deal with on computer by using simple digital circuit. The AD convert process is not necessary.

(3) The requirement of DC power is not high.

At the same time, its disadvantages are obvious too:

(1) It only can be used in the situation of single capacitance's measurement, but not

suitable for commonly used differential detection.

(2) Low precision and low stability, easy to be influenced by factors like stray capacitance and temperature.

(3) Dynamic performance is bad, and not suitable for weak capacitance detection in high frequency situation.

(4) Non-linearity of output signal is high and it needs error compensation.

2.1.6.2 Switched capacitor method

Switched capacitor method needs system time configuration to control the on-off of analog switch, and then control the charge-discharge of capacitor to be measured, to convert capacitance signal to output voltage signal. Switched capacitor method can be used either in situation of single capacitance's detection or in differential capacitance's detection.

Figure 2.15 shows the circuit diagram of detecting single capacitor using switched-capacitor method. In the Figure, C_x is the measured capacitance, C_{p1}, C_{p2}, C_{p3} and C_{p4} are the switched capacitors of switch S1, S2, S3 and S4. C_{S1} and C_{S2} are parasitic capacitance, and C is decoupling capacitance.

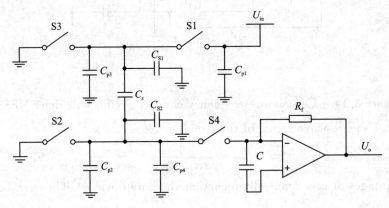

Figure 2.15 Detection circuit of switched capacitor type

If the charging frequency of the circuit is f, the output signal of the system can be expressed as

$$U_o = f \cdot U_{in} \cdot R_f \cdot C_x \tag{2.36}$$

The advantages of switched capacitance method are as follows:

(1) Output is DC signal which is proportional to measured capacitance. Demodulation of high frequency signal is not needed.

(2) The method can decrease the influence of parasitic capacitance.

Its disadvantages are as follows:

(1) Circuit structure is relatively complex.

(2) It needs complex time series circuit to control the on-off of analog switch.

(3) The charge inject effect caused by the shutting off of the analog switch will initiate the voltage peak.

(4) The continuous of charge-discharge will produce pulse noise in detection signal.

2.1.6.3 AC bridge method

AC bridge method uses capacitance to be measured consisting of the bridge or acting as one part of bridge. When the capacitance to be measured changes, the change will result in the unbalance of bridge circuit. We can amplify the signal and get output signal in direct proportion until the variation of capacitance is measured. AC bridge method is generally used in difference detection because it can use the variation of difference capacitance sufficiently.

Figure 2.16 shows the circuit diagram of one weak capacitance detecting circuit using AC bridge method[9].

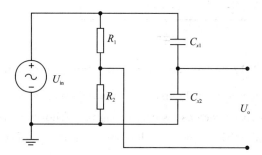

Figure 2.16 AC bridge consisting of resistors and capacitors to be measured

Suppose the difference capacitance to be measured is in initial situation $C_{x1}=C_{x2}=C$. When the variation is $\pm\Delta C$, output signal is

$$U_o = \frac{1}{2} \cdot \frac{C_{x1}-C_{x2}}{C_{x1}+C_{x2}} \cdot U_{in} = \frac{1}{2} \cdot \frac{\Delta C}{C} \cdot U_{in} \qquad (2.37)$$

The main advantages of AC bridge method are as follows:

(1) Use AC bridge structure with only a few components, and its theory is simple.

(2) Quite high sensitivity.

The main disadvantages are as follows:

(1) The amplitude of output signal is low, and the input impedance is high, so it should connect with a high input impedance operational amplifier after it.

(2) The output signal is a amplitude changeable signal, and needs subsequent circuit to demodulate it.

(3) The output signal is related to signal source, and needs AC stimulate signal source to produce signal of stable amplitude and frequency.

(4) It is easy to be influenced by stray capacitance.

2.1.6.4 Annular diode method

The schematic of the annular diode capacitance detection circuit is shown in Figure 2.17[10].

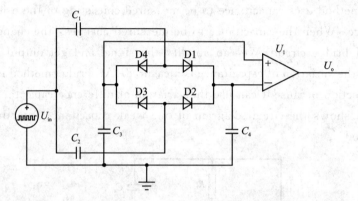

Figure 2.17 Annular diode capacitance detection circuit

This circuit shown in Figure 2.17 uses the square wave which comes from the common input of the differential capacitance as its carrier signal. Another two ends of the differential capacitance are connected with a pair of diagonal endpoints of the annular diode bridge structure which is composed by four diodes. Those two endpoints of another opposite angle are connected to the ground through two value-fixed capacitors respectively. Using a differential amplifier to get the voltage difference between those two points, we could obtain the voltage signal which is proportional to the variation of the measured capacitance.

Set the on-voltage of D1, D2, D3, D4 to V_D, the variation of the measured differential capacitance to $\pm \Delta C$ and the magnification of the differential amplifier to A. Then the output signal is

$$U_o = A \cdot (U_{C_3} - U_{C_4}) = -\frac{2 \cdot A \cdot (U_{in} - V_D)}{C_0} \cdot \Delta C \quad (2.38)$$

The advantages of the annular diode method include:

(1) The circuit is relatively simple. Only one differential amplifier is needed as the active element.

(2) The output is a DC signal proportional to the measured capacitor and does not need to demodulate the high frequency signal.

The disadvantages include:

(1) It requires high quality components. The four diodes must have good consistency.

(2) Temperature characteristics of the diode voltage drop have a great impact on the detection sensitivity.

(3) The demodulated frequency signal is amplified in the differential amplifier. Thus, the $1/f$ noise and amplifier bias voltage makes a great impact on the output signal.

2.1.6.5 Modulation and demodulation method

Modulation and demodulation method is an amplifier circuit which turns the variations of the measured capacitance into a voltage change in output. It is suitable for system simulation output.

The conversion circuit can use both integral circuit[11] and differential circuit[12].

This method has a high resolution because the noise introduced in the modulation process does not affect the post-stage circuit. Each module of the desired circuit is easy to obtain. Thus it is especially suitable for board-level circuit design (PCB).

A schematic diagram of the detection circuit based on the modem-type method is shown in Figure 2.18, which uses the integral.

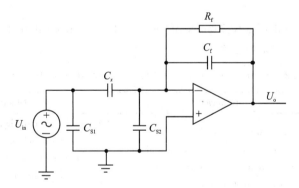

Figure 2.18 Modulation and demodulation detecting circuit based on the integration circuit

Principles of the circuit are as follows:

U_{in} is the carrier wave;

C_x is the measured capacitance;

R_f is the feedback resistor;

C_f is the feedback capacitor.

Based on the principle of virtual ground of the ideal op amp, this circuit could inhibit the stray capacitances C_{S1} and C_{S2}.

The voltage of output signal is

$$U_o = \frac{j\omega C_x R_f}{1 + j\omega C_f R_f} \cdot U_{in} \qquad (2.39)$$

When $|j\omega C_f R_f| \ll 1$, it can be expressed as

$$U_o = j\omega C_x R_f U_{in} \qquad (2.40)$$

Modulation and demodulation method has the advantages on the following aspects:

(1) It has good inhibition for the stray capacitance.

(2) Noise introduced in the modulation process does not affect the post-stage circuit. It has a high resolution.

(3) Using the high frequency carrier wave to modulated, thereby effectively avoiding the influence of $1/f$ noise.

Its disadvantages are listed as follows:

(1) The operational amplifier supply voltage is low so that the sensitivity of the detection circuit is limited.

(2) The output signal is related to the signal source so that the AC excitation signal source which is stable in amplitude and frequency is needed.

(3) The output signal is a amplitude modulated signal. It needs to be demodulated in the subsequent circuit.

The advantages and disadvantages of the weak capacitive detection methods currently used are summarized in Table 2.2.

Table 2.2 Comparison of common capacitance detection method

Methods	Advantages	Disadvantages
Capacitance frequency method	Principle is simple; Easy to get a digital signal because of the frequency output without A/D converter; Low demands on DC power	Primarily used for single capacitance detection; Stability is poor and vulnerable to the effect of tray capacitance and temperature; Large output nonlinear; Poor dynamic characteristics
Switched capacitor method	DC output signal without demodulating; Effect of the stray capacitance can be suppressed	Complex timing circuit is needed to control the switch-off; Charge injection effect of the switch will produce spikes; Continuous charging and discharging will produce impulse noise

Methods	Advantages	Disadvantages
AC bridge method	Principle is simple; High sensitivity	The effect of the stray capacitance is huge; High demands on the AC signal source; The amplitude of the output voltage is small and the output impedance is high; The output is AM signal so that the demodulation is needed
Annular diode method	Simple structure; DC output signal without demodulation	High consistency requirements of the diode; Stability is poor and vulnerable to the effect of temperature and various noise
Modulation and demodulation method	Noise introduced in the modulation process does not affect the post-stage circuit; High stability; High resolution; The effect of the stray capacitance can be suppressed	The sensitivity is limited by the power supply voltage of op amp; High requirement for AC carrier wave; The output is a amplitude modulated signal which needs to be demodulated

According to the characteristics of a capacitive MEMS device, the basic requirements for the weak capacitance detection circuit are as follows:

(1) The principle and structure are simple. In order to meet the requirements of miniaturization of MEMS devices, the principle and structure of the detection circuit should be as simple as possible.

(2) It should have a high resolution. The capacitance of the capacitor under test is very small in capacitive MEMS device. And the change in capacitance caused by the measured physical quantity is slight. Only with a high resolution could the detection circuit detect the slight change in capacitance.

(3) It has a strong anti-interference ability. Applications of MEMS devices are very extensive. Relevantly, its application environments are diverse. This situation leads to a high requirement for the stability of weak capacitance detection circuit which means that it should effectively suppress the influence of temperature, the stray capacitance and the various noise.

2.2 Relevant theoretical calculations for the MEMS cantilever beam

2.2.1 Introduction

As an important MEMS component, the MEMS cantilever beam has two main effects in MEMS as follows:

(1) It is used to support the entire MEMS device (shown in Figure 2.19). Meanwhile, the agency can be equivalent to a spring oscillator providing restoring force for the reciprocating of the movable MEMS devices.

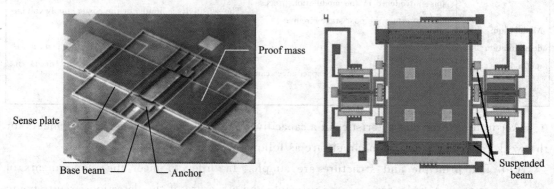

Figure 2.19　Cantilever support structure used in MEMS gyroscopes

(2) As the core component of many sensitive sensors such as MEMS force sensor (the AFM cantilever beam is shown in Figure 2.20), MEMS pressure sensor and various MEMS resonant tuning fork sensors (shown in Figure 2.21 and Figure 2.22), by detecting changes in the displacement or resonant frequency affected by the measured parameters of the cantilever beam, we could realize the measurements of measured parameters.

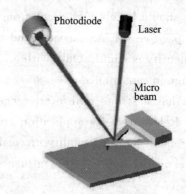

Figure 2.20　Schematic of atomic force microscopy

Chapter 2 Basic theory of MEMS

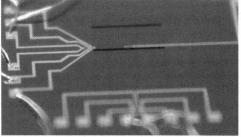

Figure 2.21 Cantilever beam structure used in resonant pressure sensor

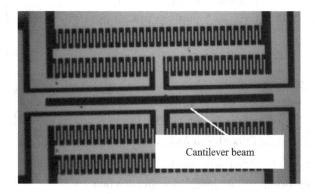

Figure 2.22 Cantilever beam structure used in tuning fork sensor

2.2.2 Theoretical calculation method for cantilever beam

Generally, as a beam, the ratio of its length and thickness is about 10:1. For cantilever beam in micro-electromechanical systems, its theories mainly involve the following aspects:

(1) The calculation of stress distribution, deformation and changes in the resonant frequency caused by changes in stress for the single-end clamped beams under the acting of axial force f.

(2) The bending deformation, stress changes and distribution of the single-end clamped beam under the effect of bending force f.

(3) The calculation of stiffness, deformation and resonant frequency for the double-end clamped beams which act as the main support structure.

2.2.3 Relevant theoretical calculation of axial tensile and compressive on single-end clamped beams

1. Calculation for the deformation of rod caused by axial tension and compression

Calculation of the relationship between those two cantilever beams under axial tension and compression could act as the theoretical basis for the design of force or pressure sensors as well as resonant beam sensors (shown in Figure 2.23).

Based on material mechanics knowledge, when the material is subjected to axial tension and compression, the relationship between axial normal strain and axial deformation can be expressed as

$$\varepsilon = \frac{\Delta L}{L} \Rightarrow \Delta L = \varepsilon L \quad (2.41)$$

The amount ε is the strain produced by the force f. L is the length of the rod. ΔL is the deformation of the rod produced by the force f. According to the generalized Hooke's law of the deformation of lever member, when

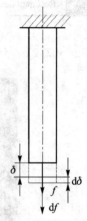

Figure 2.23 Rod under stress

the lever member is subjected to the axial force, the relationship between its strain and normal stress can be expressed as

$$\varepsilon = \frac{\sigma}{E} \quad (2.42)$$

The amount E is the elastic modulus of the material. The amount σ is the internal stress produced by the axial force f.

Similarly, under the effect of f, the relationship between the internal stress σ and the force f can be expressed as

$$f = \sigma A \quad (2.43)$$

The amount A represents the sectional area of the rod member.

When combining equations (2.41), (2.42) and (2.43), the deformation of the lever member under the axial force can be expressed as

$$\Delta L = \frac{fL}{EA} \quad (2.44)$$

The above equation can be written in the form of Hooke's law, as follow:

$$f = kx = \frac{EA}{L}\Delta L \tag{2.45}$$

The amount $\frac{EA}{L}$ indicates the axial stiffness of lever member which is dominated by the properties, shape and size of the material itself.

The corresponding strain of the lever member can be obtained from equation (2.45), as follow:

$$\varepsilon = \frac{\Delta L}{L} = \frac{F}{EA} \tag{2.46}$$

Experimental results show that: when axial stretching, the rod is axially elongated and its transverse dimension is reduced; when axial compressing, the rod is axially shortened and its transverse dimension is increased. This phenomenon come the conclusion that the transverse normal strain ε' is always an opposite sign with the axial strain ε. The experiments also show that, within the proportional limit, the transverse normal strain is proportional to the axial strain. This relationship can be expressed as

$$\mu = \left|\frac{\varepsilon'}{\varepsilon}\right| = -\frac{\varepsilon'}{\varepsilon} \tag{2.47}$$

The amount μ represents the Poisson's ratio of materials.

2. Related calculations of the rod under radial force

The stress situation of the lever member under radial force is shown in Figure 2.24.

According to the related knowledge about materials, when a force F acts on the lever member, the relationship between the amount w which is the displacement of lever member along the x direction and its corresponding torque can be expressed as

$$\frac{d^2 w}{dx^2} = \frac{M(x)}{EI} \tag{2.48}$$

E is the elastic modulus. I represents the moment of inertia of the lever member. For the rectangular cross-section, $I = \frac{bh^3}{12}$.

The rotation angle at x is

$$\theta = \frac{dw}{dx} = \int \frac{M(x)}{EI} dx + C \tag{2.49}$$

The displacement at x is

$$w = \iint \frac{M(x)}{EI} dx dx + Cx + D \tag{2.50}$$

According to the above theory, when F acts on the ends of lever member, the corresponding bending moment equation is

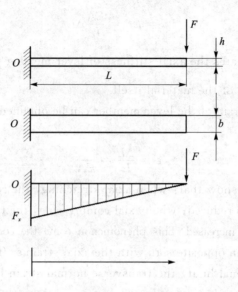

Figure 2.24 Stress situation of the lever member

$$M = F(L - x) \tag{2.51}$$

After integral operation we can get the rotation angle at x as follow:

$$\theta = \frac{dw}{dx} = \frac{F}{EI}\left(Lx - \frac{x^2}{2}\right) + C \tag{2.52}$$

We can obtain the displacement expression by integrating the equation above as

$$w = \frac{F}{EI}\left(\frac{Lx^2}{2} - \frac{x^3}{6}\right) + Cx + D \tag{2.53}$$

According to the calculation boundary conditions of the single-end clamped beams, the displacements and rotations are zero at the place where $x=0$. When taking the boundary conditions into the two equations above, we could come to the conclusion that $C=0$, $D=0$. Thus, the rotation angle and displacement can be expressed as follows:

$$\theta = \frac{dw}{dx} = \frac{F}{EI}\left(Lx - \frac{x^2}{2}\right) \tag{2.54}$$

$$w = \frac{Fx^2}{6EI}(3L - x) \tag{2.55}$$

According to equation (2.55), the largest rotation angle of the rod is at the place where $x=L$.

$$\theta = \frac{dw}{dx} = \frac{FL^2}{2EI} \tag{2.56}$$

The maximum displacement is also at the place where $x=L$.

$$w = \frac{FL^3}{3EI} \tag{2.57}$$

Thus, the rigidity of the lever member is

$$k = \frac{F}{w} = w = \frac{3EI}{L^3} \tag{2.58}$$

Similarly, based on the rod stress expression of the material mechanics, at x (confer Figure 2.24), the expression for the normal stress distribution of rods is

$$\sigma = \frac{My}{I_z} \tag{2.59}$$

where y represents the coordinate taking the neutral surface as its origin and along the direction of the interface. Then we could get the stress distribution at the cross-section of x as

$$\sigma = \frac{My}{I_z} \Rightarrow \sigma_{\max} = \frac{My}{\frac{bh^3}{12}} = \frac{12F(L-x)y}{bh^3} \tag{2.60}$$

According to Figure 2.25, the maximum stress distribution is at $\frac{y}{2}$. Bring it into the equation (2.60), then we can get that the maximum stress distribution is on its surface, and its value is

$$\sigma_{\max} = \frac{12F(L-x)}{bh^3 \cdot \frac{2}{h}} = \frac{6F(L-x)}{bh^2} \tag{2.61}$$

The maximum stress of the entire lever member along the x-direction is at the place where $x=0$:

$$\sigma_{\max} = \frac{6FL}{bh^2} \tag{2.62}$$

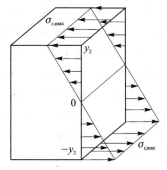

Figure 2.25 Stress distribution of beam section

From the above equation, the maximum strain and stress are at the root (clamped end) of cantilever beam, gradually decreasing along the axial direction and becoming zero at its free end. Meanwhile, appropriately increasing L/h, the ratio of the length and thickness of cantilever beam could improve the sensitivity of displacement, strain and stress on force. But it will also reduce the bending vibration frequency. For the MEMS sensor, if you want to achieve detection of a signal by way of detecting the magnitude of displacement, the end portion is a good detection position. When you want to achieve detection of stress and strain, the root is the suggested detection position.

3. Calculation for the resonant frequency of single-end clamped beam

Calculation for the resonant frequency of single-end clamped beam adopts a common equation as follow:

$$f = \frac{1}{2\pi}\sqrt{\frac{k}{m}} \tag{2.63}$$

where k represents the amount of system stiffness. The amount m represents the equivalent mass of the vibrating body. The stiffness of single-end clamped beam is $\frac{3EI}{L^3}$. And $I = \frac{bh^3}{12}$. After derivation[13], the calculation equation of resonant frequency is

$$f = \frac{0.162h}{L^2}\sqrt{\frac{E}{\rho}} \tag{2.64}$$

2.2.4 Related theoretical calculations of double-end clamped beams axial tension and compression

Double-end clamped beam and its associated theory have two main applications in MEMS sensor. First, it is used as a supporting structure for almost all of the MEMS sensors to make the movable device suspending. Its second main function is to be used as the sensitive component of resonant sensor, including single- and double-girder (such as the Double-End Tunnel Fork, DETF). In the calculation of the support structure, major concern should be given to the stiffness of structure as well as the related displacement and stress distribution caused by load changes. When it is used as the resonance sensitive element, the resonance frequency changes caused by changes in external load and the calculation of resonant frequency should be mainly concerned.

2.2.4.1 Relevant theoretical calculations of support beam

The calculation model of double-end clamped beams is shown in Figure 2.26. Generally we make the action point of the concentrated force apply to the middle position of the lever member to ensure that MEMS devices can produce the maximum displacement. Set the amount l to present the distance from the action point of the concentrated force to both ends. The driving force applied to the middle point is F.

After decoupling the force acting on C, we could get the force diagram that is shown in Figure 2.27.

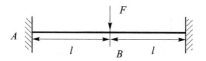

Figure 2.26 Calculation model of double-end clamped beam

Figure 2.27 The simplified model

As can be seen from Figure 2.27, point C is under the effect of concentrated force F_x and F_y which are along the directions of x and y as well as the force couple M. The boundary condition of the calculation is that the displacement of point C in direction x and y is zero after applying the driving force F. Thus, by using the Moore integral method and applying a vertical downward force on the point C, we could obtain the displacement of the point C in the vertical direction as follow:

$$\delta_y = \frac{1}{EI}\int_0^l (F_y \cdot x - M_C)x\,dx + \frac{1}{EI}\int_0^l [F_y l - M_C + (F_y - F)x] \cdot (l+x)\,dx = 0 \tag{2.65}$$

The amount E, I and l respectively express the elastic modulus of single crystal silicon material, the moment of inertia of the cantilever beam along the bending direction and the length of a single spiral arm. After simplifying equation (2.65), we get

$$16F_y l = 12M_C + 5Fl \tag{2.66}$$

Similarly, we can use the condition that the rotation angle of the cantilever is zero at the point C after applying a driving force F as another boundary condition in calculation. Applying a unit torque at point C, we could obtain the equation of rotation angle at the point C by using the Moore integral method as follow:

$$\theta_C = \frac{1}{EI}\int_0^l (F_y \cdot x - M_C)\,dx + \frac{1}{EI}\int_0^l [F_y l - M_C + (F_y - F)x]\,dx = 0 \tag{2.67}$$

After simplifying equation (2.67), we get

$$4F_y l = 4M_C + Fl \tag{2.68}$$

Combining the equations (2.66) and (2.68), we can obtain the relationships among F_y,

M_C and the driving force F as follows:

$$F_y = \frac{F}{2} \qquad (2.69)$$

$$M_C = \frac{Fl}{4} \qquad (2.70)$$

Then, applying a downward unit force on point B and utilizing the Moore integration method, we could calculate the displacement of point B under the action of the force F as

$$\delta_B = -\frac{1}{EI}\int_0^l [F_y l - M_C + (F_y - N)x] \cdot x\,dx = \frac{Fl^3}{24EI} \qquad (2.71)$$

Thus, the stiffness of the corresponding cantilever along the direction of movement can be expressed as

$$k_s = \frac{F}{\delta_B} = \frac{24EI}{l^3} \qquad (2.72)$$

2.2.4.2 Calculation of resonant frequency as the sensitive element

Most of the MEMS devices use the double-end tuning fork as the resonant sensitive component. We could solve the size of the axial force and the parameters carried by it through applying an axial force on it before and after which caused resonant frequency changes in tuning fork beam. We must consider the vibration problem of resonant beam under the impact of the axial force. This vibration phenomenon is frequently used in modeling and design of tuning fork resonator micro structure. The resonance frequency is used as a function of the design and operating of parameters, so it requires an accurate model to determine.

In practical applications, resonance tuning forks are integrated with drive unit and detection unit, as shown in Figure 2.28.

When the inertial device moves, the motion characteristics and the damping characteristics of the comb structure in micro comb-resonators are consistent. Micro comb-resonators can be simplified as the models shown in Figure 2.29.

Associated with the direction, most experts and scholars use the model shown in Figure 2.29 (a) as their study object. Only a few literatures use the model shown in Figure 2.29(b) as their

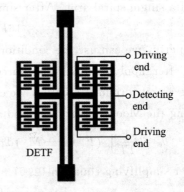

Figure 2.28 Micro comb-resonators

(a) Simplified model of pure beam (b) Simplified model of pure beam with additional proof mass

Figure 2.29 Simplified models

study object. After applying the additional mass, the resonant beam vibration model has drastically changed. Thus it is necessary to conduct a detailed derivation.

To solve this problem, we should fully consider the impact of the additional mass and axial force on the resonant tuning fork vibration characteristics and build resonant beam vibration model with additional mass. This research has a very important significance.

1. Deduction of pure tuning fork beam theory model[13]

The bending beam vibration mechanical model which only considers lateral deformation caused by bending without shearing deformation and rotary inertia effects is called the Euler-Bernoulli beam.

As shown in Figure 2.30, we analyzed the transverse free vibration of rectangular cross-section beam in the $x-y$ plane. Set $y(x,t)$ as the lateral displacement of the cross-section of the resonant beam which is x from the origin at the moment t. The amount ρ is the density of the beam. The amounts A, L, h and b are respectively cross-sectional area, length, height and width of the beam. E is the modulus of elasticity. I is the moment of inertia of the cross section of the neutral axis of the beam. Remove the micro segment $\mathrm{d}x$ at the cross section x as the research object. Its force state is shown in Figure 2.30. Using the micro segment $\mathrm{d}x$ as the research object and writing the force balance equation $\sum F_{yi} = 0$ along the y direction, we could get the following conclusion:

$$Q + \frac{\partial Q}{\partial x}\mathrm{d}x - Q - \rho A \frac{\partial^2 y}{\partial t^2}\mathrm{d}x = 0 \qquad (2.73)$$

Write down the torque balance equation $\sum M_C = 0$. It belongs to the point C which is the cross section centroid on the right side of the micro segment $\mathrm{d}x$. We can get

$$M + \frac{\partial M}{\partial x}\mathrm{d}x - M - Q\mathrm{d}x - \rho A \frac{\partial^2 y}{\partial t^2} \frac{\mathrm{d}x^2}{2} = 0 \qquad (2.74)$$

Omit the second-order small quantity dx^2.

Simplify equation (2.74) and we get

$$Q = \frac{\partial M}{\partial x} \qquad (2.75)$$

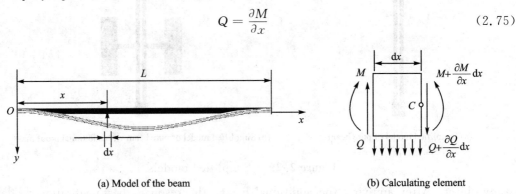

(a) Model of the beam (b) Calculating element

Figure 2.30 Transverse free vibration of beam

Bring (2.75) into equation (2.73), then we get

$$\frac{\partial^2 M}{\partial x^2} - \rho A \frac{\partial^2 y}{\partial t^2} = 0 \qquad (2.76)$$

According to material mechanics, the relationship between moment and flexural displacement is

$$M(x) = -EI \frac{\partial^2 y}{\partial x^2} \qquad (2.77)$$

Bring (2.77) into equation (2.76), then we get

$$EI \frac{\partial^4 y}{\partial x^4} + \rho A \frac{\partial^2 y}{\partial t^2} = 0 \qquad (2.78)$$

The differential equation (2.78) has the fourth-order partial derivative of spatial variable y and the second-order partial derivative of time variable t, so four boundary conditions and two initial conditions are needed to solve the equation.

Because the system has the characteristics that are time-independent and its modal is identified, the solution of equation (2.78) can be set as $y(x,t) = Y(x)T(t)$. $T(t)$ is the harmonic function.

Thus:

$$y(x,t) = Y(x)\sin(\omega t + \varphi) \qquad (2.79)$$

where ω is the vibration natural frequency of the beam. $Y(x)$ is the vibration mode function of beam. Bring equation (2.75) into equation (2.74), then we get

$$EI \frac{d^4 Y}{dx^4} - \omega^2 \rho A Y = 0 \qquad (2.80)$$

Set

$$\lambda^4 = \frac{\omega^2 \rho A}{EI} \tag{2.81}$$

Simplify equation (2.80) to

$$\frac{d^4 Y}{dx^4} - \lambda^4 Y = 0 \tag{2.82}$$

Equation (2.80) is a fourth-order homogeneous differential equation with constant coefficients. The characteristic equation is

$$s^4 - \lambda^4 = 0 \tag{2.83}$$

The four features roots of the characteristic equation (2.83) are

$$s_{1,2} = \pm i\lambda, \quad s_{3,4} = \pm \lambda \tag{2.84}$$

According to the theory of differential equations of Higher Mathematics, the general solution for equation (2.82) is

$$Y(x) = Ae^{\lambda x} + Be^{-\lambda x} + Ce^{i\lambda x} + Ae^{-i\lambda x} \tag{2.85}$$

We also know $e^{\pm \lambda x} = \text{ch}\,\lambda x \pm \text{sh}\,\lambda x$ and $e^{\pm i\lambda x} = \cos\lambda x \pm i\sin\lambda x$. Simplify equation (2.85) to

$$Y(x) = c_1 \text{ch}\,\lambda x + c_2 \text{sh}\,\lambda x + c_3 \cos\lambda x + c_4 \sin\lambda x \tag{2.86}$$

This equation is the vibration model function of Transverse free vibration of beam.

The boundary condition of the resonant beam is approximately to double sides clamped. The deflection and angle of both ends of the beam are equal to zero. There are $Y(0) = Y(L) = 0$, $Y'(0) = Y'(L) = 0$. Bring the boundary conditions into equation (2.86), we get

$$\left. \begin{array}{l} c_1 + c_3 = 0 \\ c_2 + c_4 = 0 \\ c_1 \text{ch}\,\lambda L + c_2 \text{sh}\,\lambda L + c_3 \cos\lambda L + c_4 \sin\lambda L = 0 \\ c_1 \text{sh}\,\lambda L + c_2 \text{ch}\,\lambda L - c_3 \sin\lambda L + c_4 \cos\lambda L = 0 \end{array} \right\} \tag{2.87}$$

Simplification of the above equation can be obtained as follow:

$$\left. \begin{array}{l} c_1(\text{ch}\,\lambda L - \cos\lambda L) + c_2(\text{sh}\,\lambda L - \sin\lambda L) = 0 \\ c_1(\text{sh}\,\lambda L + \sin\lambda L) + c_2(\text{ch}\,\lambda L - \cos\lambda L) = 0 \end{array} \right\} \tag{2.88}$$

For integration constants c_1 and c_2, the condition for non-zero solution is

$$\begin{vmatrix} \text{ch}\,\lambda L - \cos\lambda L & \text{sh}\,\lambda L - \sin\lambda L \\ \text{sh}\,\lambda L + \sin\lambda L & \text{ch}\,\lambda L - \cos\lambda L \end{vmatrix} = 0 \tag{2.89}$$

Because $\text{ch}^2\lambda x - \text{sh}^2\lambda x = 1$, $\cos^2\lambda x + \sin^2\lambda x = 1$, we get

$$\cos\lambda L \,\text{ch}\,\lambda L = 1 \tag{2.90}$$

Equation (2.90) is the lateral vibration frequency equation for double-end clamped beam. Solutions for the first four-order are $\lambda_1 L = 4.73, \lambda_2 L = 7.853, \lambda_3 L = 10.996, \lambda_4 L =$

14.137, so the approximate expression for the solutions is

$$\lambda_i L \approx \left(i + \frac{1}{2}\right)\pi \quad (i = 1, 2, \cdots) \tag{2.91}$$

Bring equation (2.41) into equation (2.31), then we can get the natural frequencies ω_i for all orders as

$$\omega_i = \left(\frac{2i+1}{2L}\pi\right)^2 \sqrt{\frac{EI}{\rho A}} \quad (i = 1, 2, \cdots) \tag{2.92}$$

Modal function of the first-order solution is

$$Y(x) = c_1(\text{ch}\,\lambda L - 0.9825\,\text{sh}\,\lambda L - \cos \lambda L + 0.9825\sin \lambda L) \tag{2.93}$$

Modal function of the first-order solution is the basis for the analysis of the resonant frequency.

2. Calculation for resonance frequency of resonant beam with additional mass

When there is an additional quality loaded on the resonant beam, it can be understood as increasing the maximum kinetic energy of the system from the perspective of the energy method. From the Rayleigh law, we get

$$T_{\max} = \frac{w^2}{2}\int_0^L \rho A Y(x)^2 \,\mathrm{d}x + \frac{w^2}{2} MY\left(\frac{L}{2}\right)^2 \tag{2.94}$$

$$U_{\max} = \frac{1}{2}\int_0^L EI\left[\frac{\mathrm{d}^2 Y(x)}{\mathrm{d}x^2}\right]^2 \mathrm{d}x \tag{2.95}$$

Thus the amount of angular frequency is

$$w = \sqrt{\frac{\int_0^L EI\left[\dfrac{\mathrm{d}^2 Y(x)}{\mathrm{d}x^2}\right]^2 \mathrm{d}x}{\int_0^L \rho A Y(x)^2 \,\mathrm{d}x + MY\left(\dfrac{L}{2}\right)^2}} \tag{2.96}$$

Calculate the definite integral with MATLAB, then we get

$$w = \sqrt{\frac{(500.5470 EI c_1^2)/L^3}{mc_1^2 + M \cdot 2.5222 c_1^2}} = \sqrt{\frac{198.457 EI}{(0.3965 m + M)L^3}} \tag{2.97}$$

$$m = \rho A L$$

3. Calculation for resonance frequency of resonant beam with axial force

When the resonant beam is subjected to an axial force, from the perspective of energy law, it can be understood as changing the stiffness of the resonant beam. That is changing the amount of the largest potential energy. From the Rayleigh law, we have

$$T_{\max} = \frac{w^2}{2}\int_0^L \rho A Y(x)^2 \,\mathrm{d}x \tag{2.98}$$

$$U_{\max} = \frac{1}{2}\int_0^L EI\left[\frac{\mathrm{d}^2 Y(x)}{\mathrm{d}x^2}\right]^2 \mathrm{d}x + \frac{1}{2}\int_0^L F\left[\frac{\mathrm{d}Y(x)}{\mathrm{d}x}\right]^2 \mathrm{d}x \tag{2.99}$$

Thus the amount of angular frequency is

$$w = \sqrt{\frac{\int_0^L EI \left[\frac{d^2Y(x)}{dx^2}\right]^2 dx + \int_0^L F \left[\frac{dY(x)}{dx}\right]^2 dx}{\int_0^L \rho A Y(x)^2 dx}} \qquad (2.100)$$

Calculate the definite integral with MATLAB, then we get

$$w = \sqrt{\frac{(500.547 EI c_1^2)/L^3 + (12.301\ 8 F c_1^2)/L}{m c_1^2}} = \sqrt{\frac{500.547 EI}{mL^3}\left(1 + \frac{0.024\ 58 F L^2}{EI}\right)} \qquad (2.101)$$

$$m = \rho A L$$

4. Calculation for resonance frequency of resonant beam with both additional mass and axial force

When the resonant beam is subjected to both additional mass and axial force, from the perspective of energy law, we can know that the maximum kinetic energy and the biggest potential energy change at the same time. From the Rayleigh law, we get

$$T_{max} = \frac{w^2}{2} \int_0^L \rho A Y(x)^2 dx + \frac{w^2}{2} MY\left(\frac{L}{2}\right)^2 \qquad (2.102)$$

$$U_{max} = \frac{1}{2} \int_0^L EI \left[\frac{d^2Y(x)}{dx^2}\right]^2 dx + \frac{1}{2} \int_0^L F \left[\frac{dY(x)}{dx}\right]^2 dx \qquad (2.103)$$

Thus the amount of angular frequency is

$$w = \sqrt{\frac{\int_0^L EI \left[\frac{d^2Y(x)}{dx^2}\right]^2 dx + \int_0^L F \left[\frac{dY(x)}{dx}\right]^2 dx}{\int_0^L \rho A Y(x)^2 dx + MY\left(\frac{L}{2}\right)^2}} \qquad (2.104)$$

Calculate the definite integral with MATLAB, then we get

$$w = \sqrt{\frac{(500.547 EI)/L^3 + (12.301\ 8 F)/L}{(m + 2.522\ 2) M}} = \sqrt{\frac{198.457 EI}{(0.396\ 5 m + M) L^3}\left(1 + \frac{0.024\ 58 F L^2}{EI}\right)} \qquad (2.105)$$

$$m = \rho A L$$

From a structural understanding, the DETF needs to drive and test the comb, so there must be an additional mass; from the understanding of the principle, DETF measures the amount of the axial force by measuring the change in frequency, so there must also be an axial force. Thus, equation (2.105) is the ultimately chosen equation of this thesis.

2.3 Membrane theory of MEMS

In the micromechanical pressure sensor, the diaphragm and the film are key elements for sensing the deformation caused by changes of external pressure. They also have important applications in other MEMS devices, such as fluid sensors and pumps and valves of MEMS.

In general, the diaphragm can be divided into circular diaphragm and rectangular diaphragm. This part mainly analyzes the deformation and stress distribution of diaphragm caused by changes in pressure to provide a theoretical basis for the design of related MEMS devices.

2.3.1 Theory of clamped around circular diaphragm

1. Geometric Description

The clamped around circular diaphragm is shown in Figure 2.31. R and H are the radius (m) and thickness (m) of circular flat diaphragm. E, μ and ρ are respectively the elastic modulus (Pa), the Poisson's ratio and the density (kg/m^3) of the material.

2. The boundary condition

Periphery ($r=R$) is clamped.

3. The stress state

Lower surface of the diaphragm is under the action of a uniformly distributed pressure p which can also be seen as the difference ($p_1 - p_2$) between the pressure p_2 acting on the lower surface of the diaphragm and p_1 acting on the surface of the diaphragm.

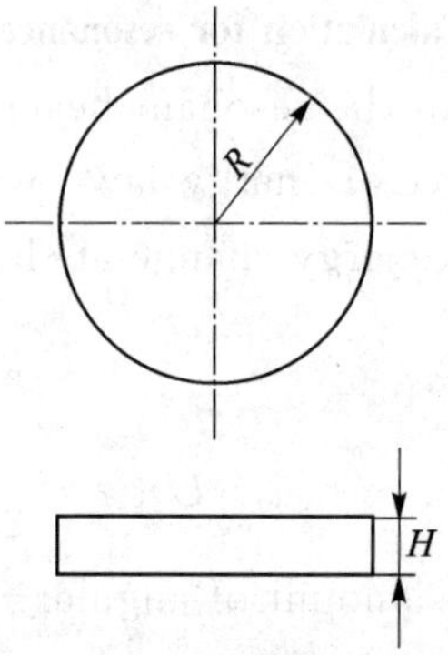

Figure 2.31 Simplified diagram for circular diaphragm

4. The basic conclusion

Under the influence of the uniform pressure p, the normal displacement of round flat diaphragm is

$$\omega(r) = \frac{\partial p(1-\mu^2)}{16EH^2}(R^2 - r^2) = \overline{W}_{R,\max}\left(1 - \frac{r^2}{R^2}\right) \qquad (2.106)$$

$$\overline{W}_{R,\max} = \frac{\partial p(1-\mu^2)}{16E} \cdot \left(\frac{R}{H}\right)^4 \qquad (2.107)$$

$\overline{W}_{R,\max}$ is the ratio of the maximum normal displacement and thickness of circular flat diaphragm.

Radial displacement of the surface of circular flat diaphragm is

$$\mu(r) = \frac{\partial p(1-\mu^2)(R^2 - r^2)r}{8EH^2} \qquad (2.108)$$

The strain and stress on the surface of the circular flat diaphragm are as follows:

$$\varepsilon_r = \frac{\partial p(1-\mu^2)(R^2-3r^2)}{8EH^2}$$
$$\varepsilon_\theta = \frac{\partial p(1-\mu^2)(R^2-r^2)}{8EH^2} \quad (2.109)$$
$$\varepsilon_{r\theta} = 0$$

$$\sigma_r = \frac{\partial p}{8H^2}[(1+\mu)R^2 - (3+\mu)r^2]$$
$$\sigma_\theta = \frac{\partial p}{8H^2}[(1+\mu)R^2 - (1+3\mu)r^2] \quad (2.110)$$
$$\varepsilon_{r\theta} = 0$$

$$f_{R.B_1}(p) = f_{R.B_1}\sqrt{1+CP} \quad (2.111)$$

$$f_{R.B_1} \approx \frac{0.469H}{R^2}\sqrt{\frac{E}{\rho(1-\mu^2)}} \quad (2.112)$$

$$C = \frac{(1+\mu)(173-73\mu)}{120}(\overline{W}_{R,\max})^2 \quad (2.113)$$

$f_{R.B_1}$ is the lowest natural frequency (Hz) of circular flat diaphragm when the pressure is zero.

C is the coefficient (Pa^{-1}) related to material, geometry parameters and physical parameters of circular flat diaphragm.

$\overline{W}_{R,\max}$ is the ratio of the maximum normal displacement and thickness at the center when the circular flat diaphragm is under a large deflection.

2.3.2 Theory of clamped around rectangular flat diaphragm

1. Geometric description

The clamped around rectangular flat diaphragm is shown in Figure 2.32. A, B, H are respectively half length (m) in x-axis, half length (m) in y-axis and thickness (m) of the rectangular flat diaphragm. Generally select x-axis as the longitudinal direction, and y-axis as its width direction. That is $A \geqslant B$. E, μ and ρ are respectively the elastic modulus (Pa), the Poisson's ratio and the density (kg/m³) of the material.

2. The boundary condition

Peripheral ($x=\pm A, y=\pm B$) is clamped.

3. The stress state

The lower surface of the diaphragm is under the action of a uniformly distributed pressure p which can also be seen as the difference (p_1-p_2) between the pressure p_2 acting

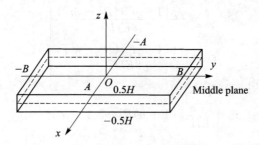

Figure 2.32 Rectangular diaphragm

on the lower surface of the diaphragm and pressure p_1 acting on the surface of the diaphragm.

4. The basic conclusion

Under the influence of uniform pressure p, the normal displacement of rectangular flat diaphragm is

$$\omega(x,y) = \overline{W}_{\text{Rec,max}} H \left(\frac{x^2}{A^2} - 1\right)^2 \left(\frac{y^2}{B^2} - 1\right)^2 \quad (2.114)$$

$\overline{W}_{\text{Rec,max}}$ is the ratio of the maximum normal displacement and the thickness of rectangular flat diaphragm.

$$\overline{W}_{\text{Rec,max}} = \frac{147 p (1 - \mu^2)}{32 \left(\frac{7}{A^4} + \frac{7}{B^4} + \frac{4}{A^2 B^2}\right) E H^4} \quad (2.115)$$

The displacements of the surface of rectangular flat diaphragm in x and y directions are as follows:

$$u(x,y) = -2 \overline{W}_{\text{Rec,max}} \frac{H^2}{A} \left(\frac{x^2}{A^2} - 1\right)^2 \left(\frac{y^2}{B^2} - 1\right)^2 \cdot \frac{x}{A}$$

$$v(x,y) = -2 \overline{W}_{\text{Rec,max}} \frac{H^2}{A} \left(\frac{x^2}{A^2} - 1\right)^2 \left(\frac{y^2}{B^2} - 1\right)^2 \cdot \frac{y}{B} \quad (2.116)$$

The strain and stress on the surface of the rectangular flat diaphragm under the uniform pressure p are as follows:

$$\varepsilon_x = -2 \overline{W}_{\text{Rec,max}} \left(\frac{H}{A}\right)^2 \left(\frac{3x^2}{A^2} - 1\right) \left(\frac{y^2}{B^2} - 1\right)^2$$

$$\varepsilon_y = -2 \overline{W}_{\text{Rec,max}} \left(\frac{H}{B}\right)^2 \left(\frac{x^2}{A^2} - 1\right)^2 \left(\frac{3y^2}{B^2} - 1\right)$$

$$\varepsilon_{xy} = -16 \overline{W}_{\text{Rec,max}} \frac{H^2}{AB} \left(\frac{x^2}{A^2} - 1\right) \left(\frac{y^2}{B^2} - 1\right) \cdot \frac{xy}{AB}$$

$$\sigma_x = \frac{-2\overline{W}_{\text{Rec,max}}E}{(1-\mu^2)}\left[\left(\frac{H}{A}\right)^2\left(\frac{3x^2}{A^2}-1\right)\left(\frac{y^2}{B^2}-1\right)^2 + \mu\left(\frac{H}{B}\right)^2\left(\frac{x^2}{A^2}-1\right)^2\left(\frac{3y^2}{B^2}-1\right)\right]$$

$$\sigma_y = \frac{-2\overline{W}_{\text{Rec,max}}E}{(1-\mu^2)}\left[\left(\frac{H}{B}\right)^2\left(\frac{3y^2}{B^2}-1\right)\left(\frac{x^2}{A^2}-1\right)^2 + \mu\left(\frac{H}{A}\right)^2\left(\frac{y^2}{B^2}-1\right)^2\left(\frac{3x^2}{A^2}-1\right)\right]$$

$$\varepsilon_{xy} = \frac{-8\overline{W}_{\text{Rec,max}}E}{1+\mu} \cdot \frac{H^2}{AB}\left(\frac{x^2}{A^2}-1\right)\left(\frac{y^2}{B^2}-1\right) \cdot \frac{xy}{AB} \tag{2.117}$$

Set $A=B$. We could get relevant conclusions about the clamped square flat diaphragm. Under the influence of uniform pressure p, the normal displacement of square flat diaphragm is

$$\omega(x,y) = \overline{W}_{S,\max} H \left(\frac{x^2}{A^2}-1\right)^2 \left(\frac{y^2}{A^2}-1\right)^2 \tag{2.118}$$

$\overline{W}_{S,\max}$ is the ratio of the maximum normal displacement and thickness of square flat diaphragm.

$$\overline{W}_{S,\max} = \frac{49p(1-\mu^2)}{192E}\left(\frac{A}{H}\right)^4 \tag{2.119}$$

The displacements of the surface of square flat diaphragm in x direction and y direction are as follows:

$$u(x,y) = \frac{-49p(1-\mu^2)}{96E}\left(\frac{A}{H}\right)^2\left(\frac{x^2}{A^2}-1\right)\left(\frac{y^2}{A^2}-1\right)^2 \cdot x$$

$$v(x,y) = \frac{-49p(1-\mu^2)}{96E}\left(\frac{A}{H}\right)^2\left(\frac{x^2}{A^2}-1\right)^2\left(\frac{y^2}{A^2}-1\right) \cdot y \tag{2.120}$$

The strain and stress on the surface of the square flat diaphragm under the uniform pressure p are as follows:

$$\varepsilon_x = \frac{-49p(1-\mu^2)}{96E}\left(\frac{A}{H}\right)^2\left(\frac{3x^2}{A^2}-1\right)\left(\frac{y^2}{A^2}-1\right)^2$$

$$\varepsilon_y = \frac{-49p(1-\mu^2)}{96E}\left(\frac{A}{H}\right)^2\left(\frac{3y^2}{A^2}-1\right)\left(\frac{x^2}{A^2}-1\right)^2$$

$$\varepsilon_{xy} = \frac{-49p(1-\mu^2)}{12E}\left(\frac{A}{H}\right)^2\left(\frac{x^2}{A^2}-1\right)\left(\frac{y^2}{A^2}-1\right) \cdot \frac{xy}{A^2} \tag{2.121}$$

$$\sigma_x = \frac{-49p}{96}\left(\frac{A}{H}\right)^2\left[\left(\frac{3x^2}{A^2}-1\right)\left(\frac{y^2}{A^2}-1\right)^2 + \mu\left(\frac{x^2}{A^2}-1\right)^2\left(\frac{3y^2}{A^2}-1\right)\right]$$

$$\sigma_y = \frac{-49p}{96}\left(\frac{A}{H}\right)^2\left[\left(\frac{3y^2}{A^2}-1\right)\left(\frac{x^2}{A^2}-1\right)^2 + \mu\left(\frac{y^2}{A^2}-1\right)^2\left(\frac{3x^2}{A^2}-1\right)\right]$$

$$\sigma_{xy} = \frac{-49p(1-\mu^2)}{24}\left(\frac{A}{H}\right)^2\left(\frac{x^2}{A^2}-1\right)\left(\frac{y^2}{A^2}-1\right) \cdot \frac{xy}{A^2} \tag{2.122}$$

References

[1] Guo Zhanshe, Feng Zhou. Theoretical and Experimental Study of Capacitance Considering Fabrication Process and Edge Effect for MEMS Comb Actuator [J]. Microsystem Technoloies, 2011, 17(1):71-76.

[2] Feng Zhou. Weak Capacitance Detection Technology on MEMS Devices and Experimental Research. Master Thesis in Beihang University. 2009,12.

[3] Huang Zhixun, Wang Xiaojin. Microwave Transmission Line Theory and Practical Techniques [M]. Beijing: Science Press, 1996: 102-104.

[4] Shi Yanling, Wu Qiang. Measuring Capacitance with Frequency Measurement Method [J]. Measurement and Equipment, 2007(2): 28-30.

[5] Liu Jun, Zhang Binzhen. Weak Signal Detection Technology. Beijing: Electronic Industry Press, 2005.

[6] Gao Xiaoding, Xu Weixing, Xu Guangshen. Research on Capacitive Force Sensor [J]. Sensor Technology, 2002,21 (7): 21-22.

[7] Yang Miaomiao, Chen Deyong, Wang Junbo. Research on Micromachined Capacitive Accelerometer Test Circuit. Sensing Technology, 2006, 19 (6): 2421-2424.

[8] Bernhard E, Boser. New Applications of Cross-Talk-Based Capacitance Measurements [C]. Proc IEEE, International Conference on Microelectronic Test Structures, 2005, 18: 257-261.

[9] Xu Tao, Song Wenai, Chen Yifang. AC Bridge Capacitive Moisture Detection Circuit Design [J]. Information Technology Aspect,2009(11): 17-20, 24.

[10] Harrison D, Dimeff J. ADiode-Quad Bridge Circuit for Use with Capacitance Transducers, Review of Scientific Instruments, 1973, 44(10): 1648-1472.

[11] Hu Hongli, Xu Tongmo, Hui Shien. A High-Accuracy, High-Speed Interface Circuit for Differential-Capacitance Transducer [J]. Sensors and Actuators, 2006, A125: 329-334.

[12] Kouji Mochizuki, Takashi Masuda, Kenzo Watanabe. An Interface Circuit for High-Accuracy Signal Processing of Differential-Capacitance Transducers [J]. IEEE Transactions on Instrumentation and Measurement, 1998, 47(4): 823-827.

[13] Fan Dajun, Liu Guangyu. New Flexible Sensor Design. Beijing: National Defense Industry Press, 1995.

[14] Fan Shangchun. Sensor Technology and Application [M]. Beijing:Beihang University Press, 2010.

Chapter 3
MEMS materials

As one of the hot research areas in the 1990s, MEMS develops with the development of related materials. For a comprehensive understanding of MEMS, in-depth study and master of materials constituting the device should be carried out. Typically, the processing of a MEMS device needs a multi-step process including growing the structure layer, the sacrificial layer and the mask layer on a substrate. Therefore, the combination of related technology and material is necessary in the study of MEMS. Etching selection ratio, material adhesion, micro-structural properties and so on should be considered while designing MEMS. MEMS devices are usually composed of a variety of materials, in which each material plays an irreplaceable role. This chapter provides a detailed description of the materials mentioned above.

3.1 Monocrystalline silicon

3.1.1 Introduction

The first silicon material used as micro sensor can be traced back to 1954. From the 1960s to the early 1970s, the mechanical and chemical micro-processing technology on a silicon substrate had continued to develop towards micro scale and strain gauges in any structure could be processed. By the mid-1970s, silicon pressure sensors had been put into commercial applications in a large scale. In the early 1980s, the all-silicon sensor began to come out after the publication of an article about the all-silicon micromechanical acceleration sensor based on monocrystalline silicon materials by Roylance and Angell from Stanford University. Since then, monocrystalline silicon MEMS materials have been widely used as the main material in MEMS. The monocrystalline silicon material plays a key role in several major aspects. As a kind of general body processing material, the monocrystalline silicon material has good properties of anisotropic etching and compatibility of the masking material. In the surface micromachining processing, monocrystalline silicon substrates are ideal structure platforms for MEMS regardless of whether the device itself is a silicon material or not. In silicon integrated MEMS devices, the monocrystalline silicon material is the primary carrier material as well. The reasons why monocrystalline silicon is so widely used are as follows:

(1) Its mechanical properties are stable, and it can be integrated into electronic devices on the same substrate, so that the input signals can be converted to electrical signals. For example, the displacement signals can be converted into electrical signals by the electrostatic comb structure integrated on silicon wafers or the piezoresistive effect.

(2) Silicon is almost an ideal construction material. It has almost the same modulus of elasticity of the steel (about 200 GPa, however it's as light as aluminium, of which the mass density is 2.3 g/cm^3). Materials with a high modulus can maintain the linear relationship between load and deformation.

(3) The melting point is 1 400 ℃, about twice that of aluminium. The high melting point allows the silicon to maintain a stable size even in the case of high temperature.

(4) Its thermal expansion coefficient is 8 times smaller than that of steel and 10 times smaller than that of aluminium, so that the impact of temperature change on related devices is extremely small, which ensures its stable performance. This feature has a great significance in improving the performance of many MEMS devices such as resonant silicon micromachined sensors.

(5) With almost no mechanical hysteresis, it's an ideal candidate for materials of sensors and brakes. Moreover, silicon is very flat, so the micro geometric structure can be formed by producing a coating or an additional film layer on it.

(6) Compared with other substrate materials, the silicon substrate has a great flexibility in design and manufacture. Processing and fabrication process of silicon substrate is comparatively mature.

Currently, the preparation of high-purity monocrystalline silicon mainly adopts the Czochralski (CZ) method. This method uses the soluble silicon method for preparation, and its basic materials are silicon oxide and silicon carbide. These materials will react and produce pure silicon and other gaseous byproducts under high temperature, and the chemical reaction is as follow:

$$SiC + SiO_2 \rightarrow Si + CO + SiO \qquad (3.1)$$

By the above reaction, the generated gas goes into the air, and the liquid silicon is left to solidify into a pure silicon. The disc-shaped pure monocrystalline silicon rods produced by this method come into three standard sizes: the diameters are 100 mm, 150 mm and 200 mm respectively. The largest size of monocrystalline silicon rod has a diameter of 300 mm, which is the latest added standard wafer size. At present, the wafer sizes and thicknesses of factory standard are as follows:

Diameter 100 mm(4 inches) ×thickness 500 μm

Diameter 150 mm(6 inches) ×thickness 750 μm

Diameter 200 mm(8 inches) ×thickness 1 mm

Diameter 300 mm(12 inches) ×thickness 750 μm

3.1.2 Crystal orientation of monocrystalline silicon

3.1.2.1 Miller indices, crystal face, crystal orientation

To understand the various properties of monocrystalline silicon, we must firstly understand some basic concepts, for example Miller indices, crystal orientation, etc.

Miller indices: Miller indices refer to relatively prime integers defined by lattice-based vector, which are used to represent the orientation of crystal face. It's also known as indices of crystal face.

In the x-y-z coordinate system, the point $P(x, y, z)$ in any plane meets the following equation:

$$\frac{x}{a} + \frac{y}{b} + \frac{z}{c} = 1 \qquad (3.2)$$

a, b and c in the above equation represent the intercepts of the plane formed by the x-axis, y-axis and z-axis respectively.

Now, set

$$\frac{1}{a} : \frac{1}{b} : \frac{1}{c} = h : k : l \qquad (3.3)$$

The values of h, k and l in equation (3.3) are Miller indices, which are the largest integers without common divisor.

If (hkl) is used to specify the plane, then use $<hkl>$ to specify the vertical direction of the plane (hkl), which refers to the crystal orientation.

So, the steps using Miller indices to determine the crystal face and crystal orientation are as follows:

(1) Identify the intercepts of a plane formed by the three axis in the rectangular coordinate system, and measure the corresponding intercepts in the unit of lattice constant.

(2) Take the reciprocals of the intercepts, and then reduce them to 3 integers without common divisor, which means reducing them to the simplest integer ratio.

(3) Use "(hkl)" to represent the result, which is the Miller index of the plane.

3.1.2.2 Calculation examples

Figure 3.1 shows the crystal structure of the monocrystalline silicon, in which the cube has 8 silicon atoms at the 8 corners. The crystal plane and crystal orientation can be expressed as follows.

1. Plane ABCD

The intercepts of the plane formed by the three axes are 1, ∞ and ∞ respectively, so $h:k:l=1:0:0$. Thus the crystal face is expressed as (100), and the crystal orientation is $<100>$ correspondingly.

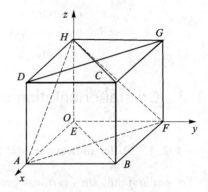

Figure 3.1 The diagram of a cube

2. Plane ADGF

The intercepts of the plane formed by the three axes are 1, 1 and ∞ respectively, so $h:k:l=1:1:0$. Thus the crystal face is expressed as (110), and the crystal orientation is $<110>$ correspondingly.

3. Plane AFH

The intercepts of the plane formed by the three axes are 1, 1 and 1 respectively, so $h:k:l=1:1:1$. Thus the crystal face is expressed as (111), and the crystal orientation is $<111>$ correspondingly.

3.1.2.3 Wafer type identification methods

Since the processing of the silicon substrate is closely related to the orientation, chip suppliers generally use the plane to show the direction in which the wafer is cut along, which is $<110>$, $<100>$ or $<111>$. The edges of the silicon monocrystalline rod can be ground into a single plane. In some cases, an additional deputy plane is given. Therefore, the wafer cutting down from these silicon rods may contain one or two planes. Generally, the main reference plane and the auxiliary reference plane are included. The main reference plane is used to specify the crystal orientation of the chip structure, while the auxiliary reference plane is used to point out the mixed type of the chip. Figure 3.2 shows the plane of P-type silicon crystal which is perpendicular to the $<111>$ direction. In this figure, the auxiliary reference plane which inclines to the main reference plane at an angle of 45° indicates that the wafer is N-type doped. Similarly, the identification of the doping type of the $<100>$ direction is shown in Figure 3.3. Silicon wafers purchased can be identified by the above method.

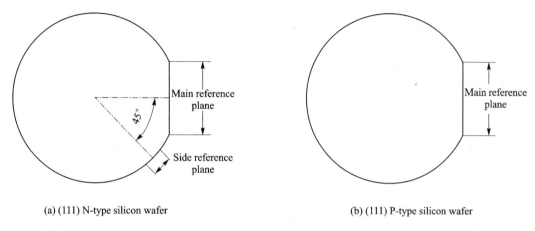

(a) (111) N-type silicon wafer   (b) (111) P-type silicon wafer

Figure 3.2 (111) as the main reference plane and the auxiliary reference plane of the silicon wafer

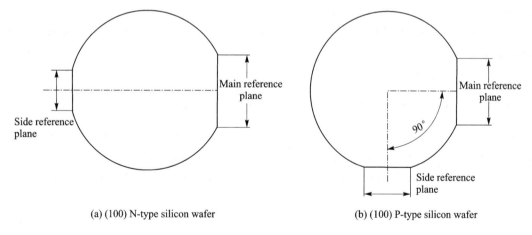

(a) (100) N-type silicon wafer   (b) (100) P-type silicon wafer

Figure 3.3 (100) as the main reference plane and the auxiliary reference plane of the silicon wafer

3.1.2.4 The crystal orientation effects on the mechanical properties of silicon wafer

The crystal orientation effect of silicon wafer makes its mechanical properties have great differences in different directions. Therefore, in the calculation of its mechanical properties, the crystal orientation of silicon wafer should be given with full consideration. Mechanical properties that the monocrystalline silicon has in different crystal orientations are shown in Table 3.1.

Table 3.1 Values of modulus of elasticity and shear that silicon crystal has in different directions

Crystal orientation of monocrystalline silicon	Modulus of elasticity E/GPa	Shear modulus G/GPa
<100>	129.5	79.0
<110>	168.0	61.7
<111>	186.5	57.5

3.1.2.5 The crystal orientation effects on the chemical properties of silicon wafer

Since the crystal lattice distances of silicon atoms that silicon wafers have in different crystal orientations are different, the chemical bonds are of different sizes ((100) is the smallest and (111) is the largest). The first result that crystal lattices in different distances lead to is that the energies needed for breaking chemical bonds are different in the chemical etching process (silicon wafer etches with the method of KOH wet), so the corrosion rate has a very big difference. Experiments show that the (100) crystal orientation has the fastest etching rate, while the (111) appears in the slowest, and different etching angles emerge during the etching process of the anisotropic wet etching of silicon.

3.1.2.6 The piezoresistive effect of monocrystalline silicon

The piezoresistive effect of silicon wafer is very common in a variety of piezoresistive sensors. Silicon piezoresistive effect is defined as follow: when the monocrystalline silicon material is subjected to external force, the resistance value is changed due to changes in the internal stress.

In 1856, Lord Kelvin discovered the piezoresistive effect of silicon wafer, referring to the effect that body resistance changes resulting from the loading machinery stress on the silicon wafer. Although a lot of materials exhibit piezoresistive properties, the piezoresistive effect of semiconductor materials which Smith discovered in 1954 has the epoch-making significance. The physical essence of piezoresistive and its relationship with the crystal orientation are established by Kanda in 1982 and Middlehoek and Audet in 1989. The piezoresistive effect of semiconductor is an order of magnitude bigger than that of metal. It is clear that monocrystalline silicon not only has good mechanical and electrical properties, but also has the excellent piezoresistive properties compared with some metal. Therefore, it becomes one of the best materials to convert mechanical deformation to electrical signals. Many MEMS devices, such as MEMS pressure sensors, piezoresistive acceleration sensors and so on, are produced by this principle. Sensors based on monocrystalline silicon piezoresistive properties are called piezoresistive sensors.

The piezoresistive effect of semiconductor materials is usually applied in two ways: one is to use the body resistances of semiconductor materials to make paste-type strain gauges; the other is to produce diffusion-type varistor or ion implanted varistor on a substrate of semiconductor material with the integrated circuit technology. This section focuses on the latter one.

The principle of sensors made by the piezoresistive effect can be expressed by resistance equation of devices. According to the definition of the size of the resistance, the change rate of resistance of any material can be written as

$$\frac{dR}{R} = \frac{d\rho}{\rho} + \frac{dL}{L} - 2\frac{dr}{r} \tag{3.4}$$

For metallic resistances, $d\rho/\rho$ is very small. Making use of the geometric deformations dL/L and dr/r to generate resistance strain effect. The device indicating the changes of resistances made by this property is called strain gauge, and the corresponding sensor is called strain sensor. However, for semiconductor materials, $d\rho/\rho$ is very big. Relatively speaking, the geometric deformations dL/L and dr/r are small enough to be neglected, which is decided by the conductive properties of semiconductor materials.

Resistance of the semiconductor material depends on the migration of a limited number of carriers, electron holes and electrons. Its resistivity can be expressed as

$$\rho \propto \frac{1}{eN_i\mu_{av}} \tag{3.5}$$

In this equation:

N_i ——carrier concentration;

μ_{av} ——the average migration rate of carrier;

e ——the electric charge, $e=1.602\times10^{-19}$ C.

When the stress acts on semiconductor materials, the number of carriers per unit volume, namely carrier concentration N_i and the average migration μ_{av} will change, and result in the changes of resistivity ρ. This is the essence of the semiconductor piezoresistive effect.

Experimental studies show that the relative change rate of resistivity of semiconductor material can be written as

$$\frac{d\rho}{\rho} = \pi_L \sigma_L \tag{3.6}$$

In this equation:

π_L——Piezoresistance coefficient(Pa^{-1}), representing the relative change of resistivity caused by a unit stress;

σ_L——Stress(Pa).

For the unidirectional stress crystal, set $\sigma_L = E\varepsilon_L$. According to equation (3.6), the change rate of resistivity can be written as

$$\frac{d\rho}{\rho} = \pi_L E \varepsilon_L \tag{3.7}$$

Take the above equation into equation (3.4), then the change rate of resistance can be written as

$$\frac{dR}{R} = \frac{d\rho}{\rho} + \frac{dL}{L} + 2\mu \frac{dL}{L} = (\pi_L E + 2\mu + 1)\varepsilon_L = K\varepsilon_L \tag{3.8}$$

$$K = \pi_L E + 2\mu + 1 \approx \pi_L E \tag{3.9}$$

The value of the elastic modulus E of semiconductor material is in the range of $3 \times 10^{11} \sim 1.9 \times 10^{11}$ Pa, and the piezoresistance coefficient π_L is in the range of $4 \times 10^{-10} \sim 8 \times 10^{-10}$ Pa, so $\pi_L E$ ranges from 50 to 150. μ represents the Poisson's ratio of the material. Therefore, in the piezoresistive effect of semiconductor materials, the strain coefficient is far greater than that of metal, which is mainly caused by the relative change of resistivity, rather than the geometric distortion. Based on the above analysis, there is

$$\frac{dR}{R} \approx \pi_L \sigma_L = \pi_L E \varepsilon_L \tag{3.10}$$

Piezoresistive sensors can be made by the piezoresistive effect of semiconductor materials, of which the main advantages are: high piezoresistance coefficient, high resolution, good dynamic response, easy to develop towards integration and intelligent. However, the maximum drawback is that the temperature coefficient of the piezoresistive effect is so big that large temperature error exists.

3.1.2.7 The determination of piezoresistance coefficient of monocrystalline silicon related to crystal orientation

For the strain effect of strain gauge, the change of the resistance value can be expressed as

$$\frac{\Delta R}{R} = K_x \varepsilon_x + K_y \varepsilon_y \qquad (3.11)$$

Accordingly, for the piezoresistive effect of semiconductor resistance, there is

$$\frac{\Delta R}{R} = \pi_a \sigma_a + \pi_n \sigma_n \qquad (3.12)$$

In this equation:

π_a, π_n—— Longitudinal and transverse piezoresistance coefficients(Pa^{-1});

σ_a, σ_n——Longitudinal(main direction) stress and transverse(vice direction) stress(Pa).

1. Piezoresistive coefficient matrix

Discuss a standard cell-dimensional cube, and it's formed along the axial directions of three standard crystal axes 1, 2 and 3(namely the x axis, y axis, z axis), which can be seen in Figure 3.4. There are three normal stresses in this micro cube: σ_{11}, σ_{22} and σ_{33}, which can be renamed to σ_1, σ_2 and σ_3. In addition there are three separate shear stresses: σ_{23}, σ_{31} and σ_{12}, which can be renamed to σ_4, σ_5 and σ_6.

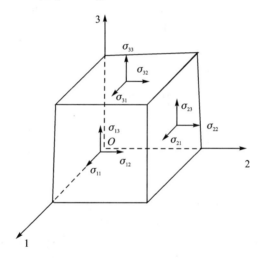

Figure 3.4 Stress distribution on monocrystalline silicon micro cube

Six independent stresses σ_1, σ_2, σ_3, σ_4, σ_5 and σ_6 will cause the relative changes δ_1, δ_2, δ_3, δ_4, δ_5 and δ_6 of six independent resistivities. Studies have shown that there is the following relationship between the stress and the relative change rates of resistance:

$$\boldsymbol{\delta} = \pi \boldsymbol{\sigma} \qquad (3.13)$$

$$\boldsymbol{\sigma}^T = [\sigma_1 \quad \sigma_2 \quad \sigma_3 \quad \sigma_4 \quad \sigma_5 \quad \sigma_6] \qquad (3.14)$$

$$\pi = \begin{bmatrix} \pi_{11} & \pi_{12} & \pi_{13} & \pi_{14} & \pi_{15} & \pi_{16} \\ \pi_{21} & \pi_{22} & \pi_{23} & \pi_{24} & \pi_{25} & \pi_{26} \\ \pi_{31} & \pi_{32} & \pi_{33} & \pi_{34} & \pi_{35} & \pi_{36} \\ \pi_{41} & \pi_{42} & \pi_{43} & \pi_{44} & \pi_{45} & \pi_{46} \\ \pi_{51} & \pi_{52} & \pi_{53} & \pi_{54} & \pi_{55} & \pi_{56} \\ \pi_{61} & \pi_{62} & \pi_{63} & \pi_{64} & \pi_{65} & \pi_{66} \end{bmatrix} \quad (3.15)$$

π is called piezoresistive coefficient matrix. It has the following characteristics:

(1) Shear stress does not lead to the positive piezoresistive effect.

(2) Normal stress does not lead to the shear piezoresistive effect.

(3) Shear stress only causes the piezoresistive effect within its own shear plane, so there is no cross-impact.

(4) There is a certain symmetry. That is: $\pi_{11} = \pi_{22} = \pi_{33}$, indicating that the positive piezoresistance effects of three directions are the same; $\pi_{12} = \pi_{21} = \pi_{13} = \pi_{31} = \pi_{23} = \pi_{32}$, indicating that the transverse piezoresistive effects are the same; $\pi_{44} = \pi_{55} = \pi_{66}$, indicating that the shear piezoresistive effects are the same.

Therefore, the piezoresistive coefficient matrix is

$$\pi = \begin{bmatrix} \pi_{11} & \pi_{12} & \pi_{12} & & & \\ \pi_{12} & \pi_{11} & \pi_{12} & & & \\ \pi_{12} & \pi_{12} & \pi_{11} & & & \\ & & & \pi_{44} & & \\ & & & & \pi_{44} & \\ & & & & & \pi_{44} \end{bmatrix} \quad (3.16)$$

There are only three separate piezoresistive coefficients, and define:

π_{11}——Longitudinal piezoresistance coefficient of monocrystalline silicon(Pa^{-1});

π_{12}——Transverse piezoresistance coefficient of monocrystalline silicon(Pa^{-1});

π_{44}——Shear piezoresistance coefficient of monocrystalline silicon(Pa^{-1}).

At room temperature, π_{11} and π_{12} of P-type silicon (hole conduction) is negligible, $\pi_{44} = 138.1 \times 10^{-11}$ Pa^{-1}. π_{44} of N-type silicon (electronic conduction) can be neglected, while π_{11} and π_{12} are bigger, and $\pi_{12} \approx -\dfrac{\pi_{11}}{2}$, $\pi_{11} = -102.2 \times 10^{-11}$ Pa^{-1}.

2. Calculation example

(1) Calculate the longitudinal and transverse piezoresistance coefficients of the crystal orientation $<010>$ in plane (001).

As shown in Figure 3.5, *ABCDEFGH* is a unit cube. *CDHG* is the plane (001), on which the crystal orientation $<010>$ is CD. The corresponding transverse is CG, namely $<100>$.

The direction cosine of $<010>$ is $l_1=0, m_1=1, n_1=0$.

The direction cosine of $<100>$ is $l_2=1, m_2=0, n_2=0$.

So

$$\pi_a = \pi_{11} - 2 \times (\pi_{11} - \pi_{12} - \pi_{44}) \times 0 = \pi_{11} \qquad (3.17)$$

$$\pi_n = \pi_{12} + (\pi_{11} - \pi_{12} - \pi_{44}) \times 0 = \pi_{12} \qquad (3.18)$$

(2) Calculate the longitudinal and transverse piezoresistance coefficients of plane (100) in the crystal orientation $<011>$.

As shown in Figure 3.6, $ABCDEFGH$ is a unit cube. $ABCD$ is the plane (100), on which the crystal orientation $<011>$ is AC. The corresponding transverse is BD.

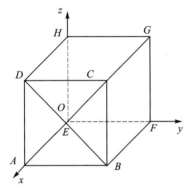
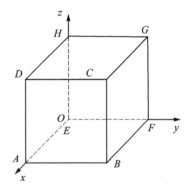

Figure 3.5 The longitudinal and transverse schematic diagram of the crystal orientation $<010>$ in plane (001)

Figure 3.6 The longitudinal and transverse schematic diagram of the crystal orientation $<011>$ in plane (100)

The vector of the direction of plane (100) can be described as i, while the vector of orientation $<011>$ can be described as $j+k$.

According to

$$i \times (j+k) = i \times j + i \times k = k + (-j) \qquad (3.19)$$

The transverse of orientation $<011>$ in plane (100) is $<0\bar{1}1>$ (usually written as $<01\bar{1}>$).

The direction cosine of $<011>$ is $l_1=0, m_1=\dfrac{1}{\sqrt{2}}, n_1=\dfrac{1}{\sqrt{2}}$.

The direction cosine of $<01\bar{1}>$ is $l_2=0, m_2=\dfrac{1}{\sqrt{2}}, n_2=\dfrac{1}{\sqrt{2}}$.

So

$$\pi_a = \pi_{11} - 2(\pi_{11} - \pi_{12} - \pi_{44})\frac{1}{2} \times \frac{1}{2} = \frac{1}{2}(\pi_{11} + \pi_{12} + \pi_{44}) \qquad (3.20)$$

$$\pi_n = \pi_{12} + (\pi_{11} - \pi_{12} - \pi_{44})\left(\frac{1}{2} \times \frac{1}{2} + \frac{1}{2} \times \frac{1}{2}\right) = \frac{1}{2}(\pi_{11} + \pi_{12} - \pi_{44}) \qquad (3.21)$$

For the P-type silicon, there are

$$\pi_a = \frac{1}{2}\pi_{44}$$

$$\pi_n = -\frac{1}{2}\pi_{44}$$

For the N-type silicon, there are

$$\pi_a = \frac{1}{4}\pi_{11}$$

$$\pi_n = \frac{1}{4}\pi_{11}$$

3.2 Polycrystalline silicon

As another kind of material which is very important in MEMS, polycrystalline silicon is generally isotropic, so there is no crystal orientation problem. Many MEMS devices are made on the base of surface micromachining of polycrystalline silicon. With this process, many complex microstructures can be produced. For example, the micro-mirror structure (shown in Figure 3.7) based on polycrystalline silicon materials produced by U.S. Sandia National Laboratory, and many acceleration sensors, gyroscopes, etc.

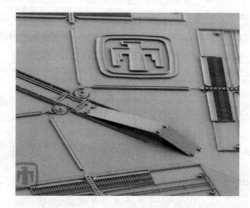

Figure 3.7 The SEM image of micro-mirror produced by polycrystalline silicon

In the application of MEMS and IC, polycrystalline silicon films are usually prepared by the process of low pressure chemical vapor deposition (LPCVD). This technology was firstly commercialized in the 1970s, and then it became a standard process of microelectronic industry. The standard polycrystalline silicon LPCVD reaction chamber is designed based on horizontal quartz tubes heated by hot wall type resistances (shown in Figure 3.8). Heating the unit quartz tube by resistance, the surface of the wafer can be heated in the furnace. Wafers are uniformly and perpendicularly placed in the quartz groove of the reaction chamber, and a certain distance is kept between them. It guarantees the form of reaction control during the deposition process, and ensures the uniform growth of film on the surface of each wafer. The so-called reaction control is that the deposition rate is determined by the reaction rate of reactants on the surface of the substrate, and the corresponding rate of diffusion control is determined by the transport rate of surface reactants. Since the deposition rate and the surface temperature are of the exponential relationship, it is necessary to precisely control the temperature of the reaction chamber. The shape retention of the reaction control motor is very good, which is very important for multi-processing.

Commonly used temperature during the deposition of polycrystalline silicon is $580 \sim 650$ ℃, and the pressure range is $100 \sim 400$ mTorr(1 Torr=133.322 4 Pa), and the most commonly used reaction gas is SiH_4, of which the decomposition equation is.

$$SiH_4 \rightarrow Si + 2H_2 \tag{3.22}$$

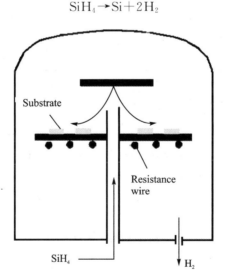

Figure 3.8 The preparation schematic of polycrystalline silicon

3.3 Silica

The silica may be thermally grown on the silicon substrate, and it can also be deposited on the silicon substrate through the technological means. As MEMS material, silica has three following functions:

(1) As the insulating layer of devices, preventing the electric conduction between devices;

(2) Acting as the material of the sacrificial layer during the surface micromachining;

(3) Acting as the mask material during the bulk silicon etching process (only one-thousandth of the corrosion rate of silicon and KOH), which is the most commonly used and main function of silica in the MEMS process.

In general, the most commonly used method for preparing the silica is the LPCVD process and the thermally oxidized process. Some mechanical properties of silica are shown in Table 3.2.

Table 3.2 Properties of silica

Properties	Values
Density/$(g \cdot cm^{-3})$	2.27
Resistivity/$(\Omega \cdot cm)$	$\geqslant 10^{16}$
Relative dielectric constant	3.9
Melting point/℃	~1 700
Specific heat capacity /$[J \cdot (g \cdot ℃)^{-1}]$	1.0
Thermal conductivity /$[W \cdot (cm \cdot ℃)^{-1}]$	0.014
Thermal expansion coefficient /$(10^{-6} \cdot ℃^{-1})$	0.5

Silicon oxidation processes can be divided into dry oxidation and wet oxidation, of which dry oxidation is introducing dry oxygen into the reactor. Wet oxidation is achieved by passing into steam, and the reaction equations are as follows:

Dry oxidation:

$$Si + O_2 \rightarrow SiO_2 \tag{3.23}$$

Wet oxidation:

$$Si + 2H_2O \rightarrow SiO_2 + 2H_2 \tag{3.24}$$

The most typical silica material is quartz material. Because of its good piezoelectric properties and a high quality factor (hundreds of thousands), it is often used as a resonator in many sensors (such as the quartz vibrating beam sensor). Specific content will be explained in piezoelectric materials, and some properties of quartz materials are shown in Table 3.3.

Table 3.3 Some properties of quartz materials

Properties	Values parallel to z	Values perpendicular to z
Thermal conductivity /[cal • ℃ • (cm • s)$^{-1}$]	29×10^{-3}	16×10^{-3}
Relative dielectric constant	4.6	4.5
Density /(kg • m^{-3})	2.66×10^3	2.66×10^3
Thermal expansion coefficient /(10^{-6} • ℃$^{-1}$)	7.1	13.2
Resistivity /(Ω • cm)	0.1×10^{15}	2×10^{16}
Fracture strength /GPa	1.7	1.7
Hardness /GPa	12	12

3.4 Piezoelectric materials

3.4.1 Piezoelectric effect and inverse piezoelectric effect of materials

For some dielectrics, when an external force is applied in a certain direction to cause the deformation of the material, internal polarization will occur, and the charge is generated on some of its surfaces. When the force is removed, it comes back to the uncharged state. The phenomenon that the mechanical energy is converted into electrical energy is called "piezoelectric effect". In turn, when applying the electric field in the dielectric polarization direction, it will generate mechanical deformation. When removing the electric field, the electrolyte deformation disappears. The phenomenon that the electrical energy is converted into mechanical energy is called "inverse piezoelectric effect", and also known as "electrostrictive effect". The "piezoelectric effect" and "inverse piezoelectric effect" are collectively known as the piezoelectric effect. Currently, small drives and even a high-frequency shaking table can be made by the use of inverse piezoelectric effect. However, piezoelectric sensors are typically achieved by utilizing the piezoelectric effect. Materials with the piezoelectric properties are known as piezoelectric materials, which can be divided into natural and synthetic piezoelectric materials. In nature, there are many kinds of piezoelectric crystals, such as quartz, potassium sodium tartrate, tourmaline, ammonium sulfate and sulfuric acid treatment, etc. Among them, the quartz crystal is a kind of natural piezoelectric material with the most practical value. Synthetic piezoelectric materials are mainly piezoelectric ceramics and piezoelectric films.

3.4.2 Quartz crystal

3.4.2.1 Piezoelectric mechanism of quartz crystal

In general, quartz crystals have regular geometry. The quartz crystal has three crystal axes, respectively x-axis, y-axis, z-axis (shown in Figure 3.9). The z-axis is the optical axis, which is determined by an optical method, without piezoelectric properties. The x-axis which is perpendicular to the optical axis through the crystal ridge is called the electrical axis. The y-axis perpendicular to plane zx is known as the mechanical axis.

Piezoelectric properties of quartz crystal are related to its internal structure. In order to illustrate its piezoelectric properties, place silicon ions and oxygen ions which compose quartz (SiO_2) crystal on the projection of plane xy perpendicular to the z-axis of crystal, and it is equivalent to the regular hexagonal arrangement in Figure 3.10. "$\oplus$" in the figure represents Si^{4+}, and "$\ominus$" represents $2O^{2-}$. When a quartz crystal is not affected by the external force, Si^{4+} and $2O^{2-}$ just distribute on the regular hexagon vertexes, forming three electric dipole moments p_1, p_2 and p_3 which are of

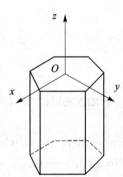

Figure 3.9 Coordinate diagram of quartz crystal

equal size and in 120° with each other, as shown in Figure 3.10(a). The value of the electric dipole moment is $p=ql$. q is the quantity of electric charge, and l is the distance between the positive and negative charges. Electric dipole moment direction is of negative charge to the positive charges. Now, the centers of positive and negative electric charges coincide, and the vector sum of electric dipole moments is equal to zero ($p_1 + p_2 + p_3 = 0$). Therefore, the crystal surface don't produce charge, and quartz crystal is electrically neutral on the whole.

When a quartz crystal is affected by the compression force along the x-axis, it causes compressive deformation along the x-axis. The relative position of positive and negative ions changes, and the centers of positive and negative charges no longer coincide, as shown in Figure 3.10(b). The component of the electric dipole moment in the direction of x-axis is

$(p_1+p_2+p_3)_x > 0$, so the crystal surface in the positive direction of x-axis appears the positive charge, and the crystal surface of the y-axis and z-axis does not occur the charge. The phenomenon of applying the force in the x-axis and generating the charge perpendicular to the crystal surface of this axis is known as the longitudinal piezoelectric effect.

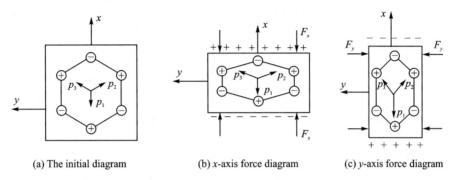

(a) The initial diagram (b) x-axis force diagram (c) y-axis force diagram

Figure 3.10 The working mechanism of quartz material

When a quartz crystal is affected by the compression force along the y-axis, it causes tensile deformation along the x-axis. The relative position of positive and negative ions changes, and the centers of positive and negative charges no longer coincide, as shown in Figure 3.10(c). The component of the electric dipole moment in the direction of x-axis is $(p_1+p_2+p_3)_x < 0$, so the crystal surface in the positive direction of x-axis appears the negative charge. Also, components in the direction of y-axis and z-axis are zero, so the crystal surface of the y-axis and z-axis does not occur the charge. The phenomenon that applying the force in the y-axis and generating the charge perpendicular to the crystal surface of the x-axis is known as the transverse piezoelectric effect.

When a quartz crystal is affected by the force along the y-axis, no matter it is the compression force or tensile force, the centers of positive and negative charges always keep a coincidence due to the equal crystal deformations in the directions of x-axis and y-axis. Components of the electric dipole moment in the directions of x-axis and y-axis are zero, so quartz crystal does not produce piezoelectric effect when applying a force along the direction of optical axis.

When the forces F_x and F_y are in opposite directions, the polarity of charge will change. Also, if a quartz crystal is affected by the same force in all directions (such as the liquid pressure), the quartz crystal will remain electrically neutral, i.e. no piezoelectric effect with volume deformation. As a kind of very important material in MEMS, piezoelectric material has very important applications.

3.4.2.2 Piezoelectric constant of quartz crystal

Take out a piece of parallelepiped from quartz crystal, and make its crystal surfaces parallel to the x-axis, y-axis and z-axis. The geometry parameters of wafer in the directions of x-axis, y-axis and z-axis are h, L, W respectively, as shown in Figure 3.11.

(1) Calculation of charge density generated on the surface perpendicular to the x-axis.

When the wafer is affected by the compression force of T_1 (N/m²) along the x-axis, it causes thickness deformation, and the charge density of σ_{11}

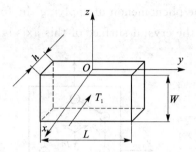

Figure 3.11 Model of quartz hexahedron

(C/m²) generated on the surface perpendicular to the x-axis is directly proportional to the stress of T_1, that is

$$\sigma_{11} = d_{11} T_1 = d_{11} \frac{F_1}{LW} \tag{3.25}$$

where d_{11} represents the piezoelectric constant, $d_{11} = 2.31 \times 10^{-12}$ C/N, referring to the charge density generated by a unit compression normal stress on the crystal surface perpendicular to x-axis when the wafer is under normal stress in the direction of x-axis. F_1 represents the compression force applied along the direction of x-axis of crystal axis (N).

The following equation is derived from equation (3.25):

$$q_{11} = \sigma_{11} LW = d_{11} F_1 \tag{3.26}$$

This equation shows that: in the above case, when the quartz wafer receives compression stress in the direction of x-axis, the electric charge q_{11} generated on the crystal surface perpendicular to the x-axis is proportional to the force F_1, and the polarity of generated charge is shown in Figure 3.12 (a). When the quartz wafer is subjected to a tensile force in the direction of x-axis, the charge is generated on the crystal surface perpendicular to x-axis, but the polarity is opposite to the compression case, as shown in Figure 3.12 (b).

When the quartz wafer is subjected to the force F_2 in the y direction, it will also generate charge on the crystal surface perpendicular to the x-axis. The charge polarity can be seen in Figure 3.12 (c) (under a compressed normal stress) or Figure 3.12 (d) (under a tensile normal stress). The relationship between the charge density and the force is

$$\sigma_{12} = d_{12} T_2 = d_{12} \frac{F_2}{hW} \tag{3.27}$$

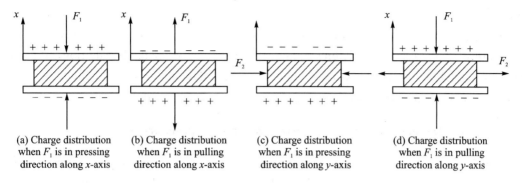

(a) Charge distribution when F_1 is in pressing direction along x-axis

(b) Charge distribution when F_1 is in pulling direction along x-axis

(c) Charge distribution when F_1 is in pressing direction along y-axis

(d) Charge distribution when F_1 is in pulling direction along y-axis

Figure 3.12 Schematic diagram of the generation mechanism of quartz crystal charge

where d_{12} represents the piezoelectric constant when the crystal bears the mechanical stress in the y direction (C/N), referring to the charge density generated on the crystal surface perpendicular to the x-axis when the wafer is under normal stress in the direction of y-axis. T_2 represents the normal stress applied along the direction of y-axis of crystal axis (N/m^2).

By equation (3.27), there is

$$q_{12} = \sigma_{12}LW = d_{12}\frac{F_2}{hW}LW = d_{12}\frac{LF_2}{h} \quad (3.28)$$

According to the axisymmetric condition of quartz crystal, there are

$$d_{12} = -d_{11} \quad (3.29)$$

$$q_{12} = -d_{11}\frac{L}{h}F_2 \quad (3.30)$$

These show that: when the force is applied along the direction of the mechanical axis to the quartz wafer, the electric charge generated on the crystal surface perpendicular to the x-axis is related to the geometric parameters of the wafer. A appropriate choice of the parameters of wafer (h, L) can increase the electric charge and improve the sensitivity.

3.4.2.3 The properties of quartz crystal

Quartz crystal is a kind of piezoelectric crystal with good performance. It requires no artificial polarization process, no pyroelectric effect, and the temperature stability of dielectric constant and piezoelectric constant is very good. In the range of 20~200 ℃, the piezoelectric constant is only reduced by 0.016% while the temperature rises by 1 ℃. When the temperature rises to 400 ℃, the piezoelectric constant d_{11} is only reduced by 5%. When the temperature rises to 500 ℃, the piezoelectric constant declines sharply. When it reaches 573 ℃ (known as the Curie temperature point), the quartz crystal will lose the piezoelectric properties.

The piezoelectric property of a quartz crystal is very stable, but relatively weak. Its temperature characteristics and long-term stability are very good. In addition, the quartz crystal material has a high nature frequency, good dynamic response, high mechanical strength, good insulation properties, good hysteresis and repeatability.

3.4.2.4 Thermal sensitivity of quartz piezoelectric resonators

Usually the relationship between the resonance frequency and the temperature of the piezoelectric resonator is called the thermal sensitivity. The thermal sensitivity of quartz piezoelectric resonators usually uses the thermal sensitivity coefficient C_t for quantitative evaluation. Thermal sensitivity coefficient C_t represents the derivation of the frequency to the temperature which is lower than a certain temperature (usually choose $t_0 = 25$ ℃), namely

$$C_t = \frac{\partial f}{\partial t}\bigg|_{t=t_0} \tag{3.31}$$

Considering the thermal sensitivity of the resonators of different working frequencies, the frequency temperature coefficient defined in equation (3.32) can be used for comparing, that is

$$T_f = \frac{C_t}{f} = \frac{1}{f}\frac{\partial f}{\partial t}\bigg|_{t=t_0} \tag{3.32}$$

The relationship between the frequency and temperature of resonators is known as the temperature-frequency characteristic. Experiments show that: the temperature-frequency characteristic of quartz resonators of any kind is a two or three times parabola, or a straight line. In the temperature range of $-200 \sim 200$ ℃, the temperature-frequency characteristic can be expressed as

$$f(t) = f_0 \left[1 + \sum_{n=1}^{3} \frac{1}{n!f_0} \frac{\partial^n f}{\partial t^n}\bigg|_{t=t_0} (t - t_0)^n \right] \tag{3.33}$$

where f_0 represents the resonant frequency (Hz) at the temperature t_0.

Equation (3.33) has a sufficient practical accuracy.

The coefficient $\frac{1}{n!f_0} \frac{\partial^n f}{\partial t^n}\bigg|_{t=t_0} \overset{\text{def}}{=} T_f^{(n)}$ is known as the n-order frequency-temperature coefficient. Among them, one-order, two-order and three-order frequency-temperature coefficient respectively are

$$T_f^{(1)} = \frac{1}{f_0}\frac{\partial f}{\partial t}\bigg|_{t=t_0} \tag{3.34}$$

$$T_f^{(2)} = \frac{1}{2f_0} \frac{\partial^2 f}{\partial t^2}\bigg|_{t=t_0} \qquad (3.35)$$

$$T_f^{(3)} = \frac{1}{6f_0} \frac{\partial^3 f}{\partial t^3}\bigg|_{t=t_0} \qquad (3.36)$$

Combining the preceding several equations, it can be expressed as

$$f(t) = f_0 \left[1 + \sum_{n=1}^{3} T_f^{(n)} (t-t_0)^n \right] = $$
$$f_0 [1 + T_f^{(1)}(t-t_0) + T_f^{(2)}(t-t_0)^2 + T_f^{(3)}(t-t_0)^3] \qquad (3.37)$$

Due to the anisotropic of quartz crystal material, its temperature coefficient is closely related to the orientation of piezoelectric element and the vibration modal adopted. This characteristic opens the way for the control of the temperature-frequency characteristic of the resonator, and also provides many possibilities for the use of the temperature-frequency characteristic of the resonator.

Based on the above analysis, for the piezoelectric quartz element which isn't sensitive to the temperature, appropriate cutting type and working mode should be chosen to reduce its frequency-temperature coefficient. But for the piezoelectric quartz sensor which is sensitive to temperature, frequency-temperature coefficient should be selected appropriately.

3.4.3 Piezoelectric ceramics

3.4.3.1 Working mechanism of piezoelectric ceramics

The piezoelectric ceramic is a synthetic polycrystalline piezoelectric material. It consists of numerous tiny electric domains. These electric domains are actually small regions of the spontaneous polarization. The direction of the spontaneous polarization is completely arbitrary arrangement. In the absence of external electric field, on the whole, polarization effects of these electric domains are offset from each other, so that the original piezoelectric ceramic is electrically neutral, and does not have piezoelectric properties.

In order to make piezoelectric ceramic with piezoelectric effect, the polarization treatment should be carried on. The so-called polarization treatment is to strengthen the electric field on the piezoelectric ceramic at a certain temperature, and the piezoelectric ceramic will have piezoelectric properties after 2~3 hours' treatment. This is because the polarization direction of electric domains in the internal ceramics tends to the direction of the electric field under the action of external electric field, and the direction is the polarization direction of piezoelectric ceramics.

The piezoelectric ceramic after the polarization treatment still remains strong internal remnant polarization when the external electric field is removed. When the piezoelectric ceramic is subjected to the external force, the electric domain boundaries move, therefore the remnant polarization will change, and the piezoelectric ceramic exhibits the piezoelectric effect.

3.4.3.2 Equivalent circuit of the piezoelectric transducer element

When the piezoelectric transducer element is subjected to external force, electric charges are generated on the two surfaces of the piezoelectric element in a certain direction (i.e. electric pole): accumulate positive charges on a surface and negative charges on another surface. So, the piezoelectric transducer element used for positive piezoelectric effect can be seen as a static charge generator. Obviously, when the two surfaces of the piezoelectric element accumulate charges, it is equivalent to a capacitor. The original capacity is

$$C_a = \frac{\varepsilon S}{\delta} = \frac{\varepsilon_r \varepsilon_0 S}{\delta} \tag{3.38}$$

In this equation,

C_a ——The capacitance of the piezoelectric element (F);

S ——The electrode surface area of the piezoelectric element (m^2);

δ ——The thickness of the piezoelectric element (m);

ε ——The dielectric constant between the plates (F/m);

ε_0 ——The dielectric constant in vacuum (F/m);

ε_r ——The relative dielectric constant between the plates, $\varepsilon_r = \varepsilon/\varepsilon_0$.

Therefore, the piezoelectric transducer element may be equivalent to the equivalent circuit of a charge source and a capacitor connected in parallel, as shown in Figure 3.13(a).

Since the following relationship exists among the open-circuit voltage u_a, the electric charge q and the capacitance C_a in the capacitor, i.e.

$$u_a = \frac{q}{C_a} \tag{3.39}$$

The piezoelectric transducer element can also be equivalent to a voltage equivalent circuit of a voltage source and a series of capacitors, as shown in Figure 3.13(b).

It's clear that the direct conversion of piezoelectric transducer element after the action by the outside world is "the electric charge" rather than "the voltage charge". In the design of the relevant circuit, the charge to voltage conversion circuit should be added.

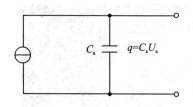
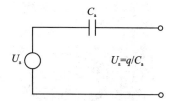

(a) Equivalent circuit under paralleling condition   (b) Equivalent circuit under series condition

Figure 3.13 The equivalent circuit of the piezoelectric element

3.5 Other MEMS materials

Other MEMS materials include silicon carbide, silicon nitride, shape-memory alloys, metallic material and polymeric materials and so on.

1. Si_3N_4 material

Si_3N_4 is widely used in MEMS, and it has good insulation, super antioxidant capacity and corrosion resistance characteristics, so it is often used as a wet etching mask material and a insulation material. There are two methods for preparing Si_3N_4, LPCVD and PECVD.

2. SiC material

Similar to the Si_3N_4 material, SiC material has good temperature stability, and it has a strong oxidation resistance even at a very high temperature. MEMS devices often deposit a layer of the film to prevent high temperature destruction. Besides, the material has a good corrosion resistance.

3. Metallic material

In MEMS, metallic material can be used as a metallic mask (such as using aluminium as the mask layer of the process of ICP (Inductively Coupled Plasma)), construction material (as shown in Figure 3.14, the micro electromagnetic motor stator coil etc.), substrate conductor material (almost all the processes adopt the sputtering method to make copper wire in the basement) and conductor material (such as the golden line by the welding method, which can be used to connect the device and test circuit).

Figure 3.15 shows the golden line in the resonant silicon micromachined gyroscope which connects to the chip by the pressure welding method). Different metals have different connection ways. Metal acted as seed layer can be generated by the sputtering or deposition method, and metal layer as construction material can be generated by the electroplating method. As the wire connected to a control circuit, it can be achieved by the pressure

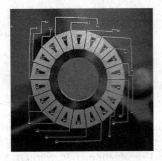 

(a) Sketch of the stator windings (b) Sketch of the enlarged windings

Figure 3.14 Micrographs of the micro electromagnetic motor stator coil

welding method. To generate the wire on the surface of the chip, the lift-off process can be used (the details will be introduced in Subsection 4.3.5).

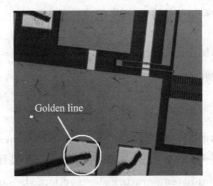

Figure 3.15 Micrograph of the connection of golden line of resonant gyroscope

3.6 Summary

This chapter introduces several commonly used MEMS materials, including monocrystalline silicon, polysilicon, silica, quartz material, piezoelectric ceramic and metallic materials and so on. Among them, monocrystalline silicon is the most commonly used MEMS material, of which the piezoresistive characteristics as well as the anisotropic caused by crystal orientation are keys to pay attention to. For quartz and piezoelectric ceramics, MEMS devices mainly use the piezoelectric effect and the inverse piezoelectric effect. The former is a natural material and the latter is for the artificial materials. At the same time, for the piezoelectric materials, the final generation is charge instead of voltage, so the charge to voltage conversion circuit should be added in the test circuit.

Chapter 4
MEMS technology

With the development of the MEMS microfabrication process, the MEMS technology develops constantly. The level of MEMS microfabrication process technology determines the performance of the device. Therefore, for the relevant professional students and researchers, it is essential to understand the working principle and the processing performance of various MEMS microfabrication technology.

Figure 4.1 is a simple fabrication process for MEMS rectangular slot. The figure shows

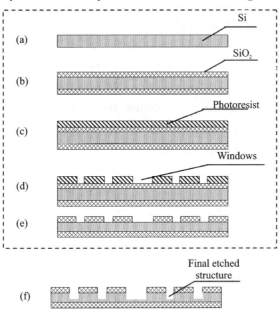

(a)—Picking of silicon; (b)—Oxidation technology; (c)—Coating of photoresist;
(d)—Lithography; (e)—HF etching of SiO_2; (f)—Etching of Silicon

Figure 4.1 A typical process flow diagram of MEMS bulk silicon

that the process mainly consists of two parts. The first part is the lithography process (Dotted line). It is used to transfer the graphics which are under engraved to the silicon chip from the photolithography. The second part is etched process. It is used to generate the required final structure (this example uses KOH solution to avoid corrosion).

This chapter will introduce the microfabrication process based on these two parts. The lithography process can be divided into wafer cleaning, oxidation, spin coating, prebaking, exposure, development, hardening, silica open window etc. In the second part, the methods of generating the final structure layer are too many to classify. They can be divided into surface silicon process, bulk silicon technology, LIGA process, quasi-LIGA process, thin film preparation process, sputtering process, bonding process and so on. The surface silicon process, bulk silicon technology, LIGA process, quasi-LIGA process, thin film preparation process and sputtering process are used to generate the structure layer, and the bonding process is used to bind different structural layers.

4.1 MEMS lithography process

Micromachining process lithography which is also known as the photolithography stems from microelectronic integrated circuit manufacturing. It is the microfabrication method which is used early in the application of micro-mechanical manufacturing field. It is still widely used and evolving now. Lithography is the key technology to manufacture MEMS devices. Its principle is similar with the photogravure in printing technology by smearing photoresist on a base material such as silicon. Then use the high-resolution energy beam to expose the photo-etch layer via mask. After developing, obtain an extremely fine geometry on the resist layer which is the same with the mask pattern. Then by etching or other methods, create miniature structure on the workpiece material.

At about 1958, the photolithography technique has been successfully applied for the first time in the manufacture of semiconductor devices. This technology has greatly promoted the invention and the rapid development of integrated circuits. For decades, integrated technology continues to miniaturization in which the lithography technology plays an important role. Till now, the width of the graphic lines is reduced by about three orders of magnitude. Now we can realize the processing of less than 1 μm line width. The integration increased by about six orders of magnitude. The integrated circuit chips contain one million or ten million components now.

The photolithographic technique is generally constructed by the basic process as follows:

(1) Original image production: In accordance with the technical requirements of product drawings, use CAD technology for processing pattern graphic design.

(2) Lithography technology to produce the master mask: Use a laser light source to exposure the photographic film and product the original image according to the NC program by a numerical control plotter.

(3) Wafer cleaning: The surface to be processed could be purified through degreasing, pickling, washing and other methods. Also, these methods make the silicon dry which is conducive to the good adhesion between the photoresist and the silicon surface.

(4) Silicon oxidation: Product a dense layer of silicon dioxide on the silicon surface through silicon oxidation process (or vapor deposition, etc.). It is the protective layer for the process behind.

(5) Spin coating process: Use a dedicated equipment to smear a layer of photoresist uniformly which is good in adhesion and considerable in thickness on the surface of the wafer under lithography.

(6) Former baking: Make the photoresist film dry to increase the adhesion of the photoresist film and the wafer surface and the wear resistance of the photoresist film. Meanwhile enable the photoresist film to have a full photochemical reaction when expose.

(7) Exposure: Cover the silicon surface which is coated with photoresist with mask or place the mask between the light source and the photoresist. Use the ultraviolet light to make the selective irradiation for the photoresist through a mask. In the place subjected to illumination, the photoresist undergoes a photochemical reaction. Thereby the nature of the photosensitive portion of the gum is changed. Accurate positioning and strict control of the exposure intensity and the exposure time is the key.

(8) Image developing: The aim of developing is to make the photoresist film on the wafer surface after exposure to present the pattern the same (positive photoresist) or opposite (negative photoresist) with the mask. To ensure the quality, strict inspections is needed for the wafer after development.

(9) Hardening: Make the adhesion between the film and the silicon tight to prevent the loss of layer and enhance the corrosion resistance of film itself.

(10) Etching: Set the photoresist after hardening process as a masking layer. Carry out the dry or wet etching process to the substrate to obtain a desired pattern.

(11) Removing of photoresist: Remove the photoresist film with a dry or wet method.

4.2 Key technology of MEMS lithography process

4.2.1 Wafer cleaning

Cleaning process of the wafer surface is the starting process of MEMS technology. It is also a very important process. Wafer cleaning is not clean (for example, dust, dirt etc.), The thickness of the photoresist is distributed unevenly in coating process. Sometimes, after the developing of photoresist, the coating surface may accumulate a lot of photoresist which is not completely developed around the impurities and form many "islands". This may directly affect the production effects and even lead to failure. Cleaning is generally done in a washing machine. First of all, in the cleaning process, select CCl_4 which is strong in decontamination capability as the cleaning agent. Then, remove the carbon tetrachloride from silicon and wash it in acetone solution, so as to achieve the purpose of removing the organic impurities from the surface of silicon. After washing with acetone, put the wafers into alcohol. This method could remove the acetone from the surface of wafer because acetone is soluble in alcohol. similarly, because alcohol was dissolved in water, wash it several times with deionized water to remove the alcohol on the surface. Finally, bake the wafer in an oven to remove the moisture on its surface. The baking temperature should not be lower than 100 ℃. To ensure the cleaning effect, the cleaning process must be carried out in a clean room with no more than one thousand dust particles per cubic meter. For surfaces which require higher surface cleanliness that the cleaning process is difficult to reach, the wafer may be placed in the cleaning fluid consisting of concentrated sulfuric acid and hydrogen peroxide (volume ratio is 1:3), and heated to 80 ℃ for cleaning. After cleaning, it can be placed directly into the deionized water to wash.

For silicon which has good surface cleanliness, we can directly rinse it in deionized water or wash it with alcohol first and then with deionized water. Finally, clean it in the oven.

4.2.2 Silicon oxidation

4.2.2.1 Theory of the thermal oxidation technology

Generally, the oxidation technology of silicon can be divided into dry oxidation technology and wet oxidation technology. Their reaction equations can be expressed as

Dry oxidation technology:

$$Si + O_2 \xrightarrow{\Delta} SiO_2 \qquad (4.1)$$

Wet oxidation technology:

$$Si + 2H_2O \xrightarrow{\Delta} SiO_2 + 2H_2 \uparrow \qquad (4.2)$$

According to Deal-Grove model[1], they have their same mathematic model:

$$y_{ox}^2 + Ay_{ox} = B(t + \tau) \qquad (4.3)$$

where A and B are constants under a certain oxidation temperature and their units are μm and $\mu m^2/h$ respectively. They are proportional to the diffusivity of the gas (O_2 or H_2O vapor); y_{ox} (μm) is the thickness of silicon dioxide; t (h) is the oxidation time; τ (h) is the compensating time, and it can be expressed as

$$\tau = \frac{y_0^2 + Ay_0}{B} \qquad (4.4)$$

where y_0 is the initial thickness of silicon dioxide film.

4.2.2.2 Computation of A and B under any oxidation temperature

A and B in equation (4.4) are very important parameters to calculate the thickness of silicon dioxide film. According to reference[1], there are just some experimental data (shown in Table 4.1). Values at the arbitrary oxidation temperatures are not mentioned. How to precisely compute A and B under any oxidation temperature is of great importance. According to reference[2], the relationship among A, $\frac{B}{A}$ and the oxidation temperature T fits the Arrhenius function, which indicates that $\lg B$ and $\lg \frac{B}{A}$ are linear with $\frac{1}{T}$. Their relationship can be expressed as

$$\lg B = a_A \frac{1}{T} + b_A \qquad (4.5)$$

$$\lg \frac{B}{A} = a_B \frac{1}{T} + b_B \qquad (4.6)$$

where a_A, b_A, a_B and b_B are constants. So according to the linear fitting method and the listed data in Table 4.1, if use dry oxidation method, their relationship can be expressed as

$$\lg B = -3828.307 \frac{1}{T} + 1.8757 \Rightarrow B = 10^{-\frac{3828.307}{T} + 1.8757} \qquad (4.7)$$

$$\lg \frac{B}{A} = -6826.952 \frac{1}{T} + 5.7103 \Rightarrow \frac{B}{A} = 10^{-\frac{6826.952}{T} + 5.7103} \qquad (4.8)$$

Table 4.1 Typical data of A and B

Temperature/°C	Dry oxidation		Wet oxidation	
	A	B	A	B
920	0.235	0.004 9	0.5	0.203
1 000	0.165	0.011 7	0.226	0.287
1 100	0.09	0.027	0.11	0.51
1 200	0.04	0.045	0.05	0.72

Table 4.2 lists the calculated relative errors between the theoretical results and experimental results.

It is concluded from the calculated results that the maximum relative error is only 7.72%. This indicates that the theoretical model at arbitrary temperature is correct.

Table 4.2 Errors between theoretical and experimental results in dry oxidation

Items / Temperature/°C	B/A			B		
	Experimental result	Theoretical result	Relative error/%	Experimental result	Theoretical result	Relative error/%
920	0.020 8	0.019 49	6.53	0.004 9	0.005 1	3.92
1 000	0.070 9	0.076 38	7.72	0.011 7	0.011 2	4.46
1 100	0.300 0	0.280 0	6.67	0.027 0	0.025 5	5.88
1 200	1.125 0	1.050 0	6.67	0.045 0	0.048 2	6.64

Therefore, A can be expressed as

$$A = \frac{B}{10^{-\frac{6\,826.952}{T}+5.710\,3}} = \frac{10^{-\frac{3.828.307}{T}+1.875\,7}}{10^{-\frac{6\,826.952}{T}+5.710\,3}} = 10^{\frac{2\,998.645}{T}-3.834\,6} \quad (4.9)$$

Correspondingly, B and $\frac{B}{A}$ in the wet oxidation technology can be expressed as

$$\lg B = -2\,231.927\,\frac{1}{T} + 1.719\,3 \Rightarrow B = 10^{-2\,231.927\frac{1}{T}+1.719\,3} \quad (4.10)$$

$$\lg \frac{B}{A} = -6\,114.352\,\frac{1}{T} + 6.237\,8 \Rightarrow \frac{B}{A} = 10^{-6\,114.352\frac{1}{T}+6.237\,8} \quad (4.11)$$

Relative errors between the theoretical results and experimental results are listed in Table 4.3.

It is concluded from the relative errors that the maximum value is 4.87%. This also indicates that the fitted result is correct in the wet oxidation technology.

Table 4.3 Error between theoretical and experimental results in wet oxidation

Items Temperature/°C	B/A			B		
	Experimental result	Theoretical result	Relative error/%	Experimental result	Theoretical result	Relative error/%
920	0.406	0.391	3.69	0.203	0.196 5	3.31
1 000	1.269 91	1.327	4.49	0.287	0.301 7	4.87
1 100	4.636 36	4.775 2	3.00	0.51	0.49	4.08
1 200	14.4	13.88	3.61	0.72	0.723 4	0.47

According to equations (4.10) and (4.11), A in the wet oxidation technology can be expressed as

$$A = \frac{B}{10^{-\frac{6\,114.352}{T}+6.237\,8}} = \frac{10^{-\frac{2\,331.927}{T}+1.719\,3}}{10^{-\frac{6\,114.352}{T}+6.237\,8}} = 10^{\frac{3\,782.425}{T}-4.517\,5} \quad (4.12)$$

4.2.2.3 Model built in the real MEMS oxidation technology for silicon

1. Oxidation method in the real MEMS oxidation technology

In MEMS technology, the oxidation film mainly acts as the shielding layer in the etching process (including the wet etching and dry etching technology). This requires the thickness of the film (1 μm or so) to be much bigger than that of IC technology. Figure 4.2 is the oxidation relationship between the oxidation time and the oxidation thicknessin of the dry oxidation technology. The oxidation temperatures are 920 °C, 1 000 °C, 1 100 °C and 1 200 °C. From the sketch, we can see that the minimum time to obtain 1 μm thickness is 23 hours for the pure oxidation though the quality of SiO_2 is good. If the heating time and cooling time are added, there is too long oxidation time for us. Too long oxidation time indicates that this method is not feasible to obtain the 1 μm SiO_2 film. Figure 4.3 shows the relationship between the oxidation time and the thickness of SiO_2 in the wet oxidation technology. It is concluded from the sketch that the minimum time to obtain the 1 μm thickness is just 1.46 hours, much less than that of the dry oxidation method. But, experiment indicates that the oxidated film with this method has poor performance to endure the etching in the bulk technology. This is also not feasible to be used in MEMS technology. So, in most engineering occasions, dry-wet-dry oxidation method is used to obtain the ideal thickness of SiO_2.

2. Model building of the dry-wet-dry oxidation technology

In this method, dry O_2 is firstly input to carry out the chemical reaction with Si. This can successfully confirm the quality of SiO_2. Then, in order to increase the oxidation speed,

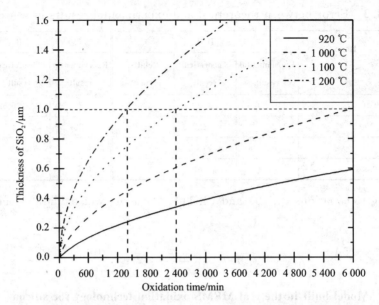

Figure 4.2 Relationship between the oxidation time and the oxidated thickness in the dry oxidation technology

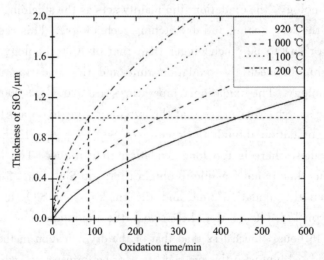

Figure 4.3 Relationship between the oxidation time and the oxidated thickness in the wet oxidation technology

water vapor is added to make the wet oxidation reaction. Finally, dry oxygen is added again to strengthen the oxidated quality. According to this description, the relationship between the thickness of SiO_2 and the oxidation time for each process can be deduced by combining equations (4.3), (4.4), (4.7),(4.9),(4.10) and (4.12) as follow:

$$y_{ox} = f(t_1, t_2, t_3) =$$

$$\left. \begin{array}{l} y_{ox1} = \dfrac{-A_1 + \sqrt{A_1^2 + 4B_1(t + \tau_1)}}{2} \quad \left(0 \leqslant t \leqslant t_1, \tau_1 = \dfrac{y_0^2 + A_1 y_0}{B_1}\right) \\[2mm] y_{ox2} = \dfrac{-A_2 + \sqrt{A_2^2 + 4B_2(t - t_1 + \tau_2)}}{2} \quad \left(t_1 \leqslant t \leqslant t_2, \tau_2 = \dfrac{y_{ox1}^2 + A_2 y_{ox1}}{B_2}\right) \\[2mm] y_{fin} = y_{ox3} = \dfrac{-A_1 + \sqrt{A_1^2 + 4B_1(t - t_2 + \tau_3)}}{2} \quad \left(t_2 \leqslant t \leqslant t_3, \tau_3 = \dfrac{y_{ox2}^2 + A_3 y_{ox2}}{B_3}\right) \end{array} \right\}$$

(4.13)

In the above equations, y_0 is the initial thickness of SiO_2 before the process; t is the oxidation time; t_1, t_2, t_3 are the end time points of the dry-wet-dry oxidation process.

$$A_1 = 10^{\frac{2\,998.645}{T_1} - 3.8346}, \qquad B_1 = 10^{-\frac{3\,828.307}{T_1} + 1.8757} \qquad (4.14)$$

$$A_2 = 10^{\frac{3\,782.425}{T_2} - 4.5175}, \qquad B_2 = 10^{-\frac{2\,231.927}{T_2} + 1.7193} \qquad (4.15)$$

$$A_3 = 10^{\frac{2\,998.645}{T_3} - 3.8346}, \qquad B_3 = 10^{-\frac{3\,828.307}{T_3} + 1.8757} \qquad (4.16)$$

In the above equations, T_1, T_2 and T_3 represent the oxidation temperature of each process.

4.2.2.4 Engineering realization of the dry-wet-dry oxidation technology

1. Engineering method for the technology

A feasible dry-wet-dry oxidation process is designed to fabricate the 1 μm SiO_2 film after so many experiments are carried to obtain the optimum process condition. This process is carried out in the oxidation oven. The pure silicon chips are sealed in a columnar quartz container (shown in Figure 4.4).

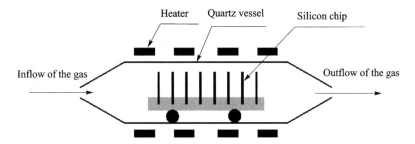

Figure 4.4 Working principle of the oxidation technology

The reaction temperature can be controlled by heat sensor that is fixed in the device. Detailed process is shown in Figure 4.5. In this figure, one grid represents 10 minutes in the horizontal axis and 10 ℃ in the vertical axis.

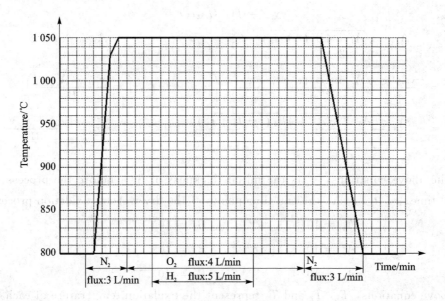

Figure 4.5 Oxidation curve

During the course of the experiment, N_2 is firstly inputted to eliminate the waste gas that may initiate reaction with silicon chip when the temperature is gradually increased. When the temperature rises to 800 ℃ for some minutes, gradually increase it to 1 050 ℃. Then, stop inputting the N_2 gas and begin to input the dry O_2 (flux: 4 L/min) to proceed the dry oxidation technology. When the dry oxidation process is carried for 30 minutes, input H_2 (flux: 5 L/min) to initiate H_2O vapor instead of directly inputting the water vapor for the convenience of operation. The reaction can be expressed as

$$2H_2 + O_2 \xrightarrow{1\,050\ ℃} 2H_2O \uparrow \qquad (4.17)$$

This reaction will not finish until H_2 is stopped inputting. Subsequently, the dry oxidation process is also stopped 60 minutes later. In order to avoid the unwanted heat reaction in the quartz oven, N_2 is still inputted into it for 60 minutes until the temperature is decreased to 800 ℃. Then, cool the silicon chip naturally.

2. Calculation of the pertinent parameters for this method

(1) Computation of A and B under 1 050 ℃

A and B under 1 050 ℃ for dry and wet oxidation can be calculated according to equations (4.14) and (4.15). The calculated results can be listed as

Dry oxidation: A—0.105, B—0.017

Wet oxidation: A—0.122, B—0.392

According to (4.13) and the listed A and B under 1 050 ℃, thicknesses of SiO_2 for each process can be listed as (μm)

$$y_{ox} = \begin{cases} 0.054 & (t = 0.5 \text{ h}, \tau_1 = 0 \text{ h}) \\ 0.840 & (t = 2.5 \text{ h}, \tau_2 = 0.06 \text{ h}) \\ 0.849 & (t = 3.5 \text{ h}, \tau_3 = 46.69 \text{ h}) \end{cases} \quad (4.18)$$

(2) Silica color judgement method

Silicon dioxide and silicon nitride layer can be distinguished by the color from the silicon substrate it grows from. The silica layer is substantially transparent and has different refractive index with the silicon substrate. The color on the surface of the silicon dioxide layer is a result of the interference of reflected light. Thus, the same color representing different oxide thicknesses is possible. Table 4.4 shows the partial list of the color on the surface of the silicon dioxide layer whose thickness is known.

Table 4.4 Color contrast of SiO_2 with different thicknesses

Thickness of SiO_2 layer/μm	0.050	0.075	0.275 or 0.465	0.310 or 0.493	0.500	0.375	0.390
Color	Sepia	Brown	Red-purple	Blue	Green-yellow or green	Green-yellow	Yellow

4.2.3 Spin coating process

Spin coating is to coat a layer of photoresist film which has good adhesion, appropriate thickness and is uniform on the surface of SiO_2. When coating, the wafers are fixed on a vacuum chuck. Drop a certain amount of photoresist in the center of the wafer. Then, rotate the silicon wafer to get a uniform layer of photoresist coating. Adhesive film layer should be uniform and has good adhesion. If the adhesive film is too thin, it may has many pinholes and poor corrosion resistance; if the adhesive film is too thick, it may has low resolution. Spin coating process is generally completed in a spin coater. The specific process can be divided into the following four steps.

(1) Pre-coated

The purpose of pre-coated is to open the photoresist on a silicon wafer. The speed of pre-coated is low, usually in hundreds of rev/min or so. In this process, approximately 65% to 85% of the solvent in photoresist will evaporate.

(2) Acceleration

Typically accelerate to several thousands of rev/min within a fraction of a second. The speed is determined by the viscosity of the photoresist (which can be classfied into thin glue and thick glue). Excess glue is left to the substrate. This step is critical for the uniformity of rotation thickness.

(3) Coating

This process forms a dry uniform resist film. The time is tens of seconds. Similarly, the length of the time is determined by the properties of the photoresist. The speed determines the final thickness of the adhesive. Time determinates the percentage content of the residual solvent. When the coating step is completed, there is still 20% to 30% of the solvent remaining.

(4) Remove the photoresist on side position

The accumulation of photoresist may happen on the edge of the wafer in the spin coating process. This results in its greater thickness. Therefore, for the photoresist which has larger viscosity, a side removing process is always required. Design the spin coating process rationally. Add a chamfer on the edge of the substrate or use the edge removing reagent to remove the side at the end of spin coating. For example, in the production process of AZ4903, fill the medical needle with acetone and push the needle in the spin coating process to fall the acetone at the edge of the wafer uniformly. The photoresist will become thin when dissolved in acetone. The edge may be thrown off the silicon wafer under the effect of centripetal force to achieve the purpose of removing the edge.

Generally, the data sheet which is incidental when purchase the photoresist will give the relationship curve (Spin Curve) between coating speed and thickness of photoresist for reference. In general, at the same rotational speed, the higher the viscosity of the glue is, the greater the thickness of the coating film will be obtained. For one kind of glue, the higher the speed of coating is, the less the thickness of the adhesive will be. But the magnitude of decrease becomes smaller and smaller when gets to a certain extent, until has nothing to do with the speed. When the coating speed is reduced, the thickness of the glue will increase, however the thickness uniformity will deteriorate. Thus we cannot reduce speed to increase the thickness of the adhesive infinitely.

1. Determination of relationship between the film thickness and the rotating speed

The determination of the relationship between the thickness of adhesive film and the rotational speed is very important in actual production process. In general, most kinds of the photoresist on market will provide a reference value of relationship between the two, so that we could obtain the desired adhesive film thickness by speed in actual production process.

Summarize several experiments data and we could get the exact relationship between the two. Use the following photoresist called AZ4903 which is high-performance and most commonly used in MEMS technology as an example to illustrate the relationship between the two.

Here is an engineering example of AZ4903 thickness determination.

Application background: Using AZ4903 to make the electromagnetic micro-motor's stator windings.

The spin coating equipment: German Karlsuss company MA6/BA6 spin coater. By programming, you can set the spin coater with the adhesive rejection time, angular velocity and angular acceleration of the turntable.

Actual operating conditions: The spin coating time is 20 seconds. The Angular acceleration is 500 rpm, and device speed will respectively be 500 rpm, 1 000 rpm, 2 000 rpm, 3 000 rpm, 4 000 rpm, and 5 000 rpm.

Preliminary determination of the thickness of the film: high-performance microscope (Olympus) which could measure the depth. First, focus on the surface of adhesive film, make it clear and read data 1; then, focus on the wafer surface, make it clear and read data 2; difference between the two is the adhesive film thickness. Then, spin coating several times and test to obtain the average of the adhesive film thickness. The results obtained are shown in Table 4.5.

Table 4.5 Speed-adhesive film thickness data table

Rotational speed/rpm	Adhesive film thickness/μm
500	35
1 000	27
2 000	22
3 000	18
4 000	13
5 000	8

2. Determination of relationship between rotational speed and thickness

According to the data shown in Table 4.5, we could obtain the relationship curve between speed and adhesive film thickness shown as curve 1 in Figure 4.6. As can be seen from the graph diagram, when the speed is greater than 1 000 rpm, the speed and adhesive film thickness are in a substantially linear relationship. When less than 1 000 rpm, thickness variation is relatively large, the two are in a non-linear relationship. Due to the large thickness corresponding to speed below 1 000 rpm, the general scope of work is greater than 1 000 rpm.

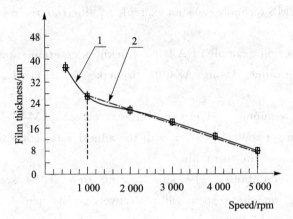

Figure 4.6 Speed-film thickness relationship curve

According to least square method, set the relationship between the rotational speed and the adhesive film thickness in the range of 1 000~5 000 rpm as

$$h = A + Bn \qquad (4.19)$$

In equation (4.19), h is the thickness of adhesive film (μm); A and B are coefficients of the linear equation.

Use the least squares method, we could calculate that $A=32.7, B=-0.005$, so the linear relationship equation of the AZ4903 adhesive rejection speed and adhesive film thickness in the range of 1 000~5 000 rpm is

$$h = 32.7 - 0.005 \qquad (1\ 000\ \text{rpm} < n < 5\ 000\ \text{rpm}) \qquad (4.20)$$

Use the linear equation (4.20), we can obtain the corresponding adhesive rejection speed according to the thickness of adhesive film. This method can be applied to other types of photoresist.

4.2.4 Prebaking

The purpose of prebaking is to let the solvent and water within the adhesive film volatile sufficiently, so that the adhesive film becomes dry. It increases the adhesion of the photoresist and silicon chip and the abrasion resistance of adhesive film. The control of prebaking time and temperature will not only affect the curing of the photoresist but also the results of the exposure and development of the photoresist. If the baking is not enough, the accuracy of exposure will deteriorates because the high solvent content of the photoresist is not sensitive to light. Exposed region cannot be completely removed in the developing time. Baking excessively may cause the photoresist to be embrittlement and reduce its adhesion. The sensitivity of the photoresist may deteriorate. The photoresist may even be carbonized and lose its original chemical properties and this may lead the exposure to failure.

The prebaking process of the photoresist is usually taken on a hot plate or oven and other special equipment. During the baking process, the temperature is generally increasing in type of Stepped. First, provide a low temperature in order to ensure that liquid solvent within the photoresist will volatile sufficiently. Then, gradually heat and ultimately reach the curing temperature. The accurate process conditions are got through several bakings and tests.

Specific examples of process conditions are shown in Figure 4.7. Because AZ4903 is a thick adhesive, to ensure the organic solvent within the photoresist to volatile sufficiently, after several tests, the 50 ℃—70 ℃—100 ℃ ladder bake method is used. After the completion of prebaking, remove the silicon from the oven and naturally cool in air. Finally get the adhesive film which has completed the prebaking and is quite satisfactory.

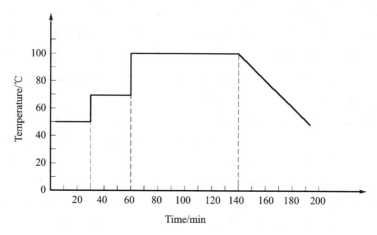

Figure 4.7 Prebaking time-temperature curve

Experiments show that, AZ4903 which has completed the prebaking in this condition, after exposure and developing, has less pinholes in the surface of adhesive film as well as good lines steepness, and almost has no glitches. No adhesive floating and photoresist shedding occur during the development.

4.2.5 Exposure

The exposure process is to transfer the graphics of MEMS devices from a lithograph to photoresist by related exposure apparatus. At present, the light source used in the exposure process may be ultraviolet light or extreme ultraviolet light. Exposure can be divided into contact exposure and non-contact exposure. Contact exposure is to contact the mask and the photoresist which is used for the production of graphics directly to exposure. The two achieve the close contact through mechanical means press, vacuum suck press and air jet compression etc. The advantages of this exposure approach are that the mask and the resist film are in close contact, the aberration is small and the resolution is improved. The disadvantage is that the close contact is easy to damage the mask and wafer.

Non-contact optical exposure technique is a method which could implement graphical copy exposure on condition that the mask and the substrate resist are not in direct contact. Since the two are separated, it will be many kinds of combinations. But the methods commonly used are proximity and projection exposure. The non-contact exposure can overcome the shortcoming that the contact exposure is easy to damage the mask and wafer. But the diffraction effect of light will reduce the resolution of pattern exposure. For non-contact exposure structure, based on diffraction effects theories, the minimum resolvable line width is $W = 15\sqrt{\lambda s/200}$ where λ is the wavelength of the irradiated light and s is the distance between the mask and the resist film.

Light is a form of electromagnetic waves which are spread in the form of fluctuations. When the light wavelength is considerable with the mask width, it will produce the diffraction effects. When light exposure or imaging on the resist through the mask pattern, the light does not propagate all in a straight line direction. Part of the light will have diffraction and go behind the mask pattern, making the edge of the lines to generate the stripes which are phase-to-phase in peak and brightness, namely the diffraction fringes. One of the characteristics of micro-machining is that the distance between the various graphic lines is close. Therefore, the phenomenon of diffraction will make the light stripe on the edge of the lines which are very close to each other coincide with each other. This will make the graphic blurred and affect the resolution.

There are two main ways to improve:

(1) Select the wavelength λ and spacing s appropriately. But s cannot be too small, otherwise it is difficult to control. And the damage to the mask and wafer are not easy to

avoid.

(2) Optimization of the light source and the optical system design.

The advantage of proximity exposure is that it extends the service life of the mask. Compared with the contact exposure method, under the same exposure times, a mask extends the service life to more than 10 times. Taking the different graphics production requirements into account, the spacing s is adjustable. In mass production, when s takes $10\sim20$ μm, we could obtain the line with $3\sim4$ μm limit width.

Factors affecting the exposure effect are mainly in the following four aspects.

1. The effect of the exposure time and intensity of exposure of the lithography results

In the lithography process, the purpose of exposure is to make the film to fully absorb the light energy and then start the photochemical reaction. The key control parameter is the amount of exposure E. The relationship between exposure intensity I_S and exposure time t_0 is[3]

$$E = I_S t_0 \qquad (4.21)$$

As can be seen from equation (4.21), the exposure amount E is proportional to the exposure intensity I_S and the exposure time t_0. When the selected exposure intensity is constant, insufficient exposure time will reduce the exposure amount, leading to insufficient light polymerization and affecting the lithography quality. Prolonged exposure can increase the effect of light diffraction and scattering. The photoresist lateral erosion is serious and resulting in low resolution. The exposure time of the photoresist is obtained by repeated experiments.

2. The impact of the contact condition of lithography mask and photoresist film on the lithography results

During the exposure, the contact condition of the photoresist mask and resist film has great impact on the resolution of the photoresist. Even if the surface of the photoresist film has a few microns of roughness, the precision lithography will decrease significantly. Furthermore, the interval between the reticle and the wafer has a significant impact on the resolution of the lithography system during the exposure. According to reference [4], the relationship between the gap G between mask plate and silicon chip and the lithography resolution (minimum line width) D is

$$G = CD^2/\lambda \qquad (4.22)$$

In this equation,

λ——Light wavelength;

C ——Coefficient associated with the system.

For a fixed exposure machine, the wavelength λ of the light selected is a fixed value. From equation (4.22), to improve the resolution of the system, the interval G between the mask plate and silicon chip must be reduced.

3. The parallelism of exposure light

The exposure light should form a parallel light beam when going through the lens and should be perpendicular to the surface of the reticle and film. Otherwise the resist pattern will be deformed or blurred. According to reference [5], during exposure the relational expression of the line width W and the exposure light for the incident angle θ of photomask is

$$W = K \frac{\lambda}{NA\left(1 + \sigma + \frac{\sin\theta}{NA}\right)} \tag{4.23}$$

where

$$\sigma + \frac{\sin\theta}{NA} < 1$$

and NA is the numerical aperture; λ is the wavelength of the outgoing light; σ is the correlation coefficient for the space segment. θ is the angle of incidence of the exposure light to the reticle. According to equation (4.23), for a fixed lithography machine, NA, λ and σ are substantially fixed. Therefore, to reduce the width of the error, must try to make $\sin\theta$ close to 1. So it should try to ensure that the exposure light perpendicular to the surface of the film to avoid distortion and blur of lithography graphics.

4. Quality of the mask

The film thickness of thick rubber process is much larger than the film thickness of the IC process, thus require a longer exposure time in the lithographic process. The mask plate which is produced in accordance with the requirements of IC production process cannot meet the requirements of photolithography. Ways to improve are to increase the thickness of the chrome (chrome on the mask), and to reduce the damage of the photoresist which should be protected. In addition, the precision of the lines of the mask will have a direct impact on the lithographic effect.

4.2.6 Development

The purpose of developing is to dissolve the photoresist (positive resist) of the photosensitive portion, leaving the portion of film which is not photosensitive, thereby obtaining the desired pattern. For a certain high performance photoresist, generally there is a special developing solution for processing. The main conditions of the control of development are the concentration of the developing solution, the developing temperature and the developing time. Configure the developing stock solution (developer with no water distribution) with deionized water according to volume ratio for the developing process. Different photoresist has different requirements on volume ratio. The specific values of volume ratio, the developing time and other conditions are also obtained by the summary of several experiments. For example, in use of photoresist AZ4903, under the conditions that the film thickness is 35 microns, the exposure time is 220 seconds, the exposure power is 300 watts, use the ways of 1:1, 1:2, 2:3, 1:3 and 1:4 to configure its developer solution and do the comparison experiments.

After repeated comparison of the experimental results, the result obtained under the microscope is listed in Table 4.6. The phenomenon shown in this table shows that 1:2 and 2:3 volume ratios are better configuration than the others. Therefore, finally select 1:2 as the developer ratio during developing. Using this ratio, we get a good experimental result which is shown in Figure 4.8.

Table 4.6 Comparison of experimental results

Shadow fluid: water (Volume ratio)	Development time/min	Phenomena and results
1:1	2'15"	In about two minutes the developing is completed. During the developing process it can be clearly seen that the red photoresist is dissolved on both the remaining portion and the removing portion of the surface of photoresist. Therefore, the loss of the part which is needed to retain is also relatively large.
2:3	3'10"	The development has slowed and the effect is more obvious. However, under the microscope, there still exist some serrate on the line edge of the developing gum.
1:2	4'10"	Speed control is relatively easy. After development the damage on the part of photoresist which is needed to retain is relatively small, lines are clear, steepness is good and the serrated on the edge of lines is small. It is a appropriate ratio.
1:3	3'	After three minutes developing the portion which is to be removed has almost no change. The developing is very slow, so this proportion of the developer should be given up.

Shadow fluid: water (Volume ratio)	Development time/min	Phenomena and results
1:4	No	Taking the fact that the volume ratio obtained by this method is smaller than the former, so give up the use of developer at this ratio.

Figure 4.8 Partial graphic of the stator windings after lithography

Problems which usually arise in development process are as follows.

(1) Developing shortage. Line is wider than the normal one, and there is a slope on the side or the photoresist on the substrate is not removed clearly. This phenomenon may be due to lack of exposure or response time, or the low concentration of the developing solution.

(2) Over developing. Removing too much photoresist in developing process will cause the narrow and incomplete graphics. Possibly due to overexposure, the developing concentration is too high or the development time is too long.

4.2.7 Hardening

Like the previous drying process, hardening process is also a heat treatment step. That is, at a certain temperature, baking the silicon wafer after development. The photoresist has softened and expanded after developing. The adhesion between the adhesive film and the surface of wafer may decrease. If put the adhesive film directly into the etching solution, the vacuum phenomenon due to etching solution will occur and this phenomenon will affect the results of production. The purpose of harden process is to let the residual photoresist solvent as well as the moisture entering during the developing process volatile. Improve the adhesion between the photoresist and silicon chip and increase the corrosion resistance of photoresist. This enables the photoresist to truly play a protective role and get ready for the next etching process. This also removes the residual developing solution. Hardening process is also done in a hot plate or in an oven. Its temperature will be slightly lower than the prebake temperature. Specific hardbaking time and temperature are obtained through repeated experiments and optimization based on data provided by the manufacturer.

4.2.8 Fabrication of the SiO$_2$ window

In the process of micro-structure fabrication which is using the MEMS technology, because the resist itself has the color problem and other issues, the process could easily lead to contamination of the structure. Thus, in general, SiO$_2$ is the material generally used on the mask layer which is used in the process of making the structure. Its production uses the oxidation process. Therefore, before etching, we should open a SiO$_2$ window. The specific process is shown in Figure 4.9. The process uses a hydrofluoric acid solution to remove SiO$_2$ material.

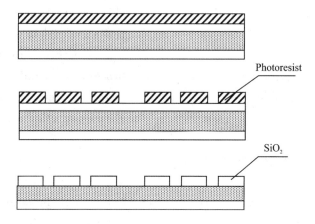

Figure 4.9 Process of opening the SiO$_2$ window

This utilizes the complexation reaction between the hydrofluoric acid and silica. The reaction equation is shown as follow[6]:

$$\begin{array}{r} SiO_2 + 4HF = SiF_4 + 2H_2O \\ \underline{SiF_4 + 2HF = H_2[SiF_6]} \\ SiO_2 + 6HF = H_2[SiF_6] + 2H_2O \end{array} \quad (4.24)$$

Pure hydrofluoric acid can easily penetrate the photoresist layer and continue to underetch from the bottom. This causes unglued. So usually add right amount of ammonium fluoride as the buffering agents in hydrofluoric acid etching solution. This method avoids the photoresist flaking so that the corrosion can be carried out smoothly. This hydrofluoric acid solution which has buffering agents is traditionally called the hydrofluoric acid buffer. The ratio of hydrofluoric acid, ammonium fluoride and water is

$$\text{Hydrofluoric acid:ammonium fluoride:deionized water} = 3 \text{ (ml):} 6 \text{ (g):} 10 \text{ (ml)} \quad (4.25)$$

The concentration of hydrofluoric acid is 48%. The corrosion rate is not only related to the ratio of the three but also the corrosion temperature and density of silica. In general, the corrosion temperature can be selected between 30 ℃ and 40 ℃. The completed silica window is shown in Figure 4.9.

4.3　Subsequent process of MEMS

MEMS subsequent process primarily refers the related processes which can proceed on the basis of lithography, including bulk silicon process, surface silicon technology, LIGA process, sputtering process, bonding process, lift-off process etc. Bulk silicon technology mainly refers to the silicon etching process, including wet etching and dry etching etc; the surface silicon technology refers to the process of making microstructure using the sacrificial layer (commonly use silicon dioxide as the material of sacrificial layer) technology. LIGA process is mainly used for the mass production of the structure with high aspect ratio or metal structures. Sputtering process and bonding process are mainly used as a secondary process of the three processes above to make the more complex microstructure; Lift-off process is mainly used to produce metal wire of the device and other structures.

4.3.1　Bulk silicon technology

Bulk silicon technology refers to the production process of the needed three-dimensional micro-structures through the corrosion of base material (usually it is monocrystalline silicon). Based on the different ways of the removal of silicon, it can be divided into wet etching and dry etching.

4.3.1.1　Wet etching process

1. KOH wet etching principles and process conditions

As one of the important MEMS technologies, the wet etching process has the advantages of simple equipment, mass production and low cost so that it is widely used. Its simplest procedure is to put the silicon wafer which is completed in the opening of SiO_2 window (shown in Figure 4.9) in the etching solution with a certain concentration. Currently, the most commonly used etching solution in wet etching process is the KOH solution. This method has the advantages of simple manufacturing process, low cost, controllable conditions etc. Therefore, for devices without high requirements on device size, aspect ratio or steepness, this method can meet the needs on the production of structure. In actual project, after repeated groping, the process conditions that the KOH solution concentration is 40% and the temperature of etching is 80 ℃ are selected. Stir the etching solution constantly in corrosion process (can be realized with a magnetic stirrer) in order to ensure the concentration of each part of the solution is the same. Thereby, the various parts have a same corrosion rate. The KOH wet etching process reaction equation is shown as follow:

$$Si + 2KOH + H_2O \xrightarrow{\Delta} K_2SiO_3 + 2H_2 \uparrow \qquad (4.26)$$

In the actual etching process, in order to increase the reaction rate with the silicon chip, the KOH solution would be added with a certain percentage of isopropanol. The KOH solution can also react with the silicon dioxide mask on the surface, and the reaction equation is

$$SiO_2 + 4KOH \xrightarrow{\Delta} K_4SiO_4 + 2H_2O \qquad (4.27)$$

Although the potassium hydroxide can react with the silicon and silicon dioxide simultaneously, however, due to the rate of the reaction with silicon is much larger than the rate of the reaction with silicon dioxide, therefore, the silicon dioxide mask can still play a good role.

Before the start of wet etching, in order to obtain the desired pattern, use the hydrofluoric acid etching solution on the silicon dioxide to carry out the part etching and obtain the desired window (shown in Figure 4.9).

2. The effect of crystal orientation on wet etching process

We know from Chapter 2, because the monocrystalline silicon wafers has different crystal orientations, the wet etching of silicon wafers belongs to anisotropic etching. For single crystal silicon with different crystal orientations, van der Waals force between its two atoms has different magnitude. This leads to the fact that during the reaction the energy needed on the destruction of the chemical bonds is different. Therefore, the corrosion rates along different crystallographic orientation are different during the corrosion process. KOH etchs the slowest on the (111) plane, and is 1/400 of the etching rate on the (100) plane.

Thus the etching rate perpendicular to (111) crystal plane is very low. In most cases it is negligible, so consider the (111) crystal plane to be the stop surface of KOH etching.

The reason why etch rates are different on different crystal planes is not yet fully clear. Generally believe that it is related to the bond density of crystal planes. The larger the density of molecule on crystal plane is and the smaller the distance between molecules is, the greater the number of connection key and their intensity will be and the higher the bond density will be. The energy required for a chemical reaction is high, so the etching speed is low.

The angles between the (111) crystal plane and the surface on the (100) silicon wafer is different from that on the (110) silicon wafer, thus the resulting structures after etching are different. The angle between the (111) plane and the surface on the (100) silicon wafer is 54.74°. Therefore, when the etching window is rectangular and parallel to the trimming of silicon wafers, the etching structure of the (100) plane is a inverted trapezoid surrounded by four (111) planes. The angle between the (111) plane and the (100) surface is 54.74°. The underside of the trapezoidal is located on silicon wafer surface and shrinks with the decrease of depth. If the etching time is long enough and the wafer thickness is sufficient, the four inclined (111) planes will gradually shrink and intersect, and finally form an inverted triangular pyramid (rectangular mask opening) or a shape of pyramid (square mask opening).

The resulting corrosion angle is shown in Figure 4.10.

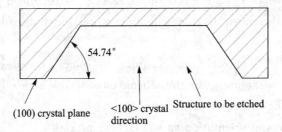

Figure 4.10 Influence of crystal orientation on corrosion

It can be seen that, compared to other forms of corrosion, the wet etching has the advantages of low cost, simple process and easy implementation etc. However, the problem of crystal orientation makes it difficult to ensure the quality of corrosion (for example, the steepness requirement of lines and processing size requirements etc.). Therefore, for situations requiring high device steepness and size and so on, the dry etching process is often used.

4.3.1.2 Dry etching process

Dry etching process uses the gases as the corrosive rather than the liquid chemicals to remove base material. During the reaction the gas mainly exists in the presence of a plasma state. It is etched by the plasma. Since the method uses gas as its corrosive, the crystal orientation of single-crystal silicon material will not affect its performance. Therefore, with this method, although its cost is high and the process is complex, we can realize the structure production with high steepness requirement and have high dimensional accuracy. The typical process most commonly used in dry etching are reactive ion etching and deep reactive ion etching. For the single-crystal silicon, the most commonly used etching gas is SF_6.

Currently, the most widely used plasma etching technique is the ICP (Inductively Coupled Plasma) etching technique. In China, almost all MEMS devices which have certain requirements on the performance of size are based on this process. This process will ionize the gas while it is going through the high voltage electric field and generate plasma. After converged in magnetic field, the plasma will bombard the wafer surface to carry out selective chemical corrosion reaction with silicon. Then generate gas which can be taken away by flow.

Figure 4.11 is a schematic diagram of the basic principle of ICP deep etching. During the experiment the main gas put into the vacuum chamber is SF_6 and O_2 (SF_6 has the corrosive effect; O_2 has the passivation effect to prevent the sidewall from corrosion). The SF_6 generates the high-speed motion plasma group through electron impact. The ionization equation is

$$SF_6 + e^- \rightarrow S_xF_y^+ + S_xF_y^* + F^* + e^- \qquad (4.28)$$

The symbol * on the upper right corner of $S_xF_y^*$ and F^* in the above equation denotes the chemically active groups. It is capable to carry out the selective chemical reaction with Si on the etching layer.

The reaction equation of F^* and wafer is as follow:

$$Si + 4F^* \rightarrow SiF_4 \uparrow \qquad (4.29)$$

Finally, the SiF_4 gas generated is released to complete the corrosion of silicon. Because the material generated in the reaction with silicon wafer is gas, the structure obtained by the method is relatively clean. Thus ensure the cleanliness of the material surface. Figure 4.12 shows a SEM (Scanning Electronic Microscopy) chart of a comb drive produced by this process. As can be seen from this figure, this process protects the performance of the device

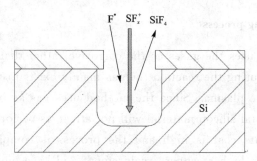

Figure 4.11 The principle of ICP process

Figure 4.12 SEM image of the comb drive in testing institutions

to be well processed.

Compared to other etching techniques, ICP process has the advantages of high corrosion depth, dimensional accuracy, aspect ratio and so on. However, because the ICP etching uses the reaction of gas and silicon wafer to obtain the desired structure, in production process, the gas ratio, reaction temperature and time selection have a large effect on production. If not managed properly, it may lead to the occurrence of over-etching and serious lateral erosion or other phenomenon, resulting in the failure of production.

Figure 4.13 is a SEM picture of structure generated due to the severe side etching. As can be seen from this figure, the long thin beam of the structure has a serious over-etching. This leads to the brokenness of the beams and the collapse of rigid block or other phenomena in testing process. Therefore, the optimum conditions are required to obtain by repeated experiments. The requirements for relevant personnel are relatively higher.

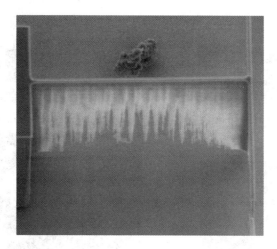

Figure 4.13 SEM image of ICP over-etching

4.3.2 Surface silicon process

Typical silicon surface micromachining process is shown in Figure 4.14. This method mainly use sacrificial layer technology to achieve a three-dimensional movable structure on the silicon surface. Sacrificial layer technology is also called separation layer technology. By using chemical vapor deposition method to form micro parts on silicon substrate, insert the separation layer material on the surrounding voids of members (mainly SiO_2). Finally, the process of dissolving or etching to remove the separation layer and separate micro components and substrates, can also be used to manufacture micromechanical links with the substrate slightly such as electrostatic micro-motor, micro gear, crankshaft and micro bridging of the vibration sensor.

Ideal sacrificial layer material must meet the requirements of the design, and the film thickness must be in an acceptable tolerance. Uneven deposition will lead to surface roughness or irregularities. When the gap is very small ($\leqslant 5$ μm), parameters are particularly important. When disengaged, the sacrificial layer must be removed integrally. The etch selectivity and the etch rate of the sacrificial layer must be high to make the structure of the other parts will not be significantly damaged. However in the actual sacrificial layer etching, a narrow groove that is hundreds of microns typically takes quite a long time. Generally, in the process of selecting silicon dioxide as sacrificial layer material, its removal mainly use the previously described hydrofluoric acid etchant.

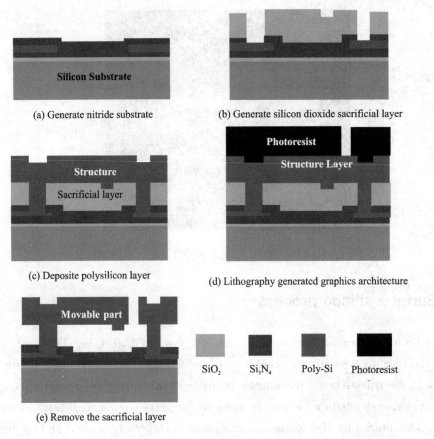

Figure 4.14 Typical silicon surface process

4.3.3 LIGA technology

LIGA (abbreviation of German "lithograph galvanformung und abformug") is the method of fabrication of high aspect ratio structures, which is invented by German Karlsruhe Research Center in 1985. The process mainly includes three parts: lithography, electroplating and casting. This is the process of using X-ray thick rubber, high-energy X-ray generator and synchronous plating equipment batch manufacturing metal and plastic structure of high aspect ratio. The process flow diagram is shown in Figure 4.15. The basic process sequence is: coating thick photoresist on a conductive substrate (its thickness varies from a few microns to several millimeters), and after X-ray exposure and development get the three-dimensional structure of the photoresist; then, use conductive layer as a plating seed layer and X-ray photoresist as the plating mold; then, the copper, gold, nickel or nickel alloy is plated in the cavity by electroplating technology; finally, remove the three-dimensional structure, and this three-dimensional metal structure can either be used as the final device needed or precision casting plastic molds, casting or pressing making a lot of plastic devices identical to the photoresist structure. The substrate LIGA processed must be electrically conductive or depositing a conductive layer on insulator substrate. LIGA can create a variety of polymer materials, metals and ceramic materials like PZT, PMNT, aluminium oxide, zirconium oxide.

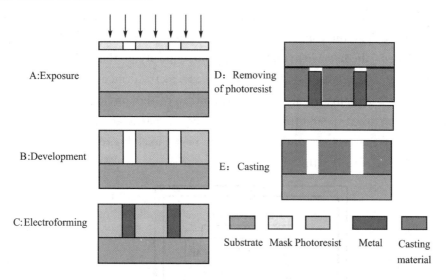

Figure 4.15　LIGA process diagram

4.3.4　Sputtering technology

Sputtering technology by using the method of high-energy particles bombarding the target makes the atom of the target from chemically bound to become a free particle, and then deposite at a certain velocity to the surface of a substrate, forming a thin film seed layer. If we want to get the seed layer, first of all we must have a target same as preparing deposited seed layer. Then, there must be a sufficiently high voltage electric field. Finally, in order to prevent the material from reacting with the gas in air during bombardment process, the process must be carried out in vacuum. Figure 4.16 shows the simplified schematic of the silicon surface layer of sputtered copper seed layer. In this figure, sputtering copper primarily used as an electrical seed layer, can be combined with lift-off process to produce MEMS wire or combined with electroforming process making three-dimensional metal structure layer. The process is carried out in the SP-I type 2 kW RF matching unit. In the high vacuum chamber of the machine (to prevent oxidation of copper at high temperatures) are two high voltage electrodes. Before going through high voltage, introduce required inert gas (argon) into the vacuum chamber. After high voltage passing, the inert gas under high electric field effect discharge, and produce a large amount of ions. These ions are accelerated by an electric field and formed high-energy ion flow, bombarded to the surface of copper target. Since the ion kinetic energy over the copper target binds energy of atoms and molecules, the copper atoms in the target escape out and sputter to the anode (oxidation finished silicon) with high speed and deposite a thin film and complete the sputtering process. As a result, the desired seed electroformed layer is formed and its thickness is about 2 000 angstroms. One disadvantage of this process is that the rate of seed layer depositing is low. In addition, the high-energy particles will produce radiation to human body during exercise, so in the process the best choice for relevant staff is to leave the scene to avoid impacts on health.

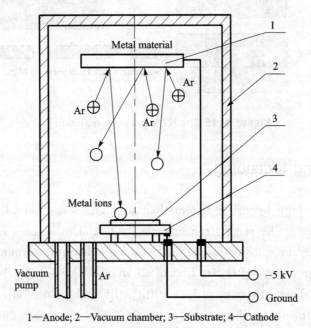

1—Anode; 2—Vacuum chamber; 3—Substrate; 4—Cathode

Figure 4.16 Sputtering schematic

4.3.5 Lift-off process

Lift-off process in MEMS technology often combines with sputtering process to generate the desired structure such as metal wire or the electroformed seed layer three-dimensional metal structures used in the production. Lift-off process has a very important role in the MEMS process. In general, the seed layer material in the lift-off process is metal. The process diagram of the lift-off process is shown in Figure 4.17. Firstly, through a lithography process, the manufactured pattern identical to desired metal structure on substrate is fabricated. Then, by a sputtering process, a layer of metal material is deposited on the surface to act as the seed layer. After the completion of the deposition, the entire wafer is placed in an organic solvent. So the dissolve photoresist (such as acetone solution) and metal seed layer sputtering on the photoresist are dissolved. The seed layer material contacted to the base material (mostly silicon) is left. The final structure is the needed metal wire. Experiment indicates that this method producing wire structure is full compliance with the requirements of MEMS devices.

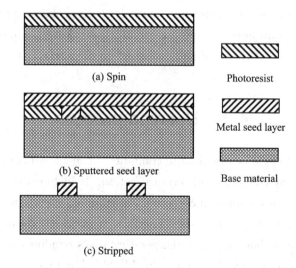

Figure 4.17 Process diagram of lift-off process

4.4 Film preparation technology

Film preparation technology refers to the course of fabricating a layer of film material on the substrate using the so-called Physical Vapor Deposition (PVD) or Chemical Vapor Deposition (CVD) method, etc. Thickness of the film is in the range from some nanometers to tens of micrometers. There are many kinds of film materials. According to the using purpose, they can be divided into metal, semiconductor silicon, germanium, glass insulators, ceramics etc. From the conductive consideration, it could be metal, insulator or superconductor. From structural considerations, it could be monocrystalline, polycrystalline or amorphous superlattice material. From the chemical composition considerations, it could be simple substance, compound or inorganic material, organic material etc.

There are many ways for the preparation of the film, they can be summed up as follows:

(1) Gas phase film formation methods, consisting of CVD such as heat, light or plasma, and PVD such as vacuum evaporation, sputtering, ion plating, molecular beam epitaxy or ion implantation filming.

(2) Liquid phase film formation methods, including the chemical plating, electroplating, dip coating etc.

(3) Other film formation methods, including spraying, coating, rolling, printing, extruding and so on.

Among them, CVD means utilizing thermal energy, glow discharge plasma, UV irradiation, laser irradiation energy alone or combined, making gaseous substance chemical reaction on the hot surface of the solid and deposited on the surface, thereby forming a stable film of solid material.

4.5 Bonding process

Wafer bonding technique is a method to combine silicon and silicon, silicon and glass, or other materials closely by chemical and physical action. It includes the connection of silicon and silicon, silicon and glass, glass and ceramic, silicon and metal, metal and metal, and it also includes the overall package of sensor. There are many methods such as anodic bonding, hot melt bonding, eutectic bonding, low temperature glass bonding, pressure welding, laser welding and electron beam welding etc. In MEMS bonding process, connecting and packaging should meet the following technical requirements:

(1) Thermal residual stress must be as small as possible;

(2) Mechanical decoupling must be made to prevent the interference of external stress;

(3) Sufficient mechanical strength and sealing ability must be strong enough (including a vacuum seal);

(4) Good electrical insulation properties must be obtained.

4.5.1 Anodic bonding process

Anodic bonding, also known as electrostatic bonding or field-assisted bonding, was put forward by Wallis and Pomerantz in 1969. Anodic bonding can bond silicon and glass, alloy and the semiconductor, metal and alloys together in the electrostatic field without any intermediate adhesive. This bonding temperature is low, and the bonding interface is solid, and has good air tightness and long-term stability, so anodic bonding is widely used. The process may be carried out in air or vacuum. Specific environmental requirements are based on the actual project. The principles of process equipment is shown in Figure 4.18 (silicon-glass bonding). Supply wafers with positive power, supply glass with negative power, and the voltage is between 200 V and 1 000 V. Heat the glass-silicon wafer to 300~500 ℃. When the voltage is applied, Na^+ ion in glass will drift to negative.

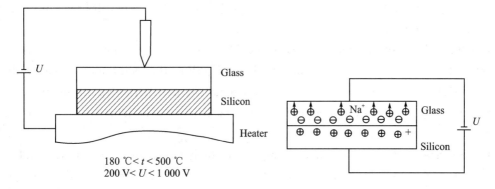

Figure 4.18 Anodic bonding process schematics

The depletion layer, which is about some micrometers in thickness, will be formed on the surface. Depletion layer is negatively charged, while silicon is positively charged. There is a large electrostatic attraction between the silicon and glass, so that the two closely contact. Thus the applied voltage is mainly applied to the depletion layer. The change of current in the circuit can reflect the electrostatic bonding process. After the voltage is applied, there is a large current pulse, then the current is reduced, and finally almost zero, indicating that the bonding has been completed at this time. The electrostatic attractive force plays a very important role in electrostatic bonding. For example, after the bonding is completed and the sample is cooled to room temperature, the charge depletion layer does not completely disappear. The remaining charge induces mirror positive charge in the silicon, and the electrostatic force between them is about 1 MPa, so the small residual charge can still produce substantial bonding force. Furtherly, at a relatively high temperature, the chemical reaction will occur in the interface of closely contact silicon-glass. This will form a strong chemical bond such as Si-O-Si bond. After several tests, following phenomena are founded:

(1) The key to make silicon-glass interface electrostatic bonding firmly and stably is the interface has sufficient Si-O bond formation.

(2) After the reverse voltage is applied at a high temperature, silicon-glass interface electrostatic bonding is still firm and stable.

(3) After the failure of electrostatic bonding, glass can be used again when the reverse voltage is applied on it.

There are many factors affecting the electrostatic bonding, including:

(1) Thermal expansion coefficients of the two electrostatic bonding materials should approximately match. Otherwise, after the bonding is completed, it will break in the cooling process due to the large internal stress.

(2) The shape of the anode affects the bonding effect. Commonly used anodes are parallel plate electrode and point-contact electrode. For the point-contact electrodes the bonding interface does not produce voids, while for the double parallel plate electrodes, there will be some parts of pores on bonding body interface, and the bonding rate is faster than the former.

(3) Surface conditions also have an impact on the bonding force. The flatter and cleaner the bonding surface is, the better the bonding quality will be. The larger the surface relief is, the smaller the electrostatic attraction will be.

The voltage upper limit of electrostatic bonding is that the glass does not breakdown, while the lower limit is the ability to cause the bonding material elastic, plastic or viscous flow and deformation. When silicon-glass is bonded, the thickness of oxide layer on silicon generally is less than 0.5 mm. Electrostatic bonding technique can also be applied to bonding between metal and glass, FeNiCO alloy and glass, metal and ceramics and so on.

4.5.2 Silicon-silicon direct bonding

In silicon-silicon bonding process, two wafers can directly bond together by heat treatment without any binder or mezzanine between them, and they do not need applied electric field either. This bonding technology is called silicon direct bonding (SDB) technology. In this technique, the silicon wafer may be heated to above 1 000 ℃, to make it in a molten state. Molecular force makes the wafers bond together. It is also called silicon fusion bonding. Its simplified process is shown in Figure 4.19. The bonding process requires the surface of two wafers having good surface roughness.

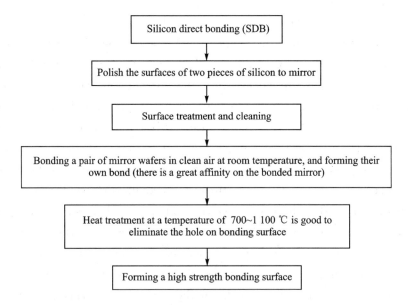

Figure 4.19 The simplified process of silicon bonded principle

It can be achieved by surface polishing process. Specific steps are listed as follows:

(1) Use hydrofluoric acid solution to soak two polished silicon wafers (either oxidation or oxidative will be ok).

(2) The polished surface of two wafers are bonded together at room temperature.

(3) Put the bonded wafers in an oxygen or nitrogen atmosphere at a high temperature for several hours to form a good bond.

Direct bonding process is simple. Bonding mechanism can be described in three stages as follows:

(1) In the first stage, when working temperature is up to 200 ℃, two silicon surfaces absorb OH groups, generating hydrogen in contacting region. When temperature is 200 ℃, polymerization reaction happened between the surfaces of two wafers. Water and the silicon-oxygen bond are formed at the same time.

$$Si-OH+HO-Si \rightarrow Si-O-Si+H_2O \tag{4.30}$$

When the temperature is up to 400 ℃, polymerization reaction is substantially completed.

(2) In the second stage, the temperature is in the range of 500~800 ℃. The diffusion from water to SiO_2 is not obvious during the course of the formation of the silicon-oxygen bond. OH group can undermine a key of bridging oxygen atom and turn it into a non-bridging oxygen atom. That is

$$HOH + Si-O-Si = 2H + 2Si-O \qquad (4.31)$$

(3) In the third stage, when the temperatures is above 800 ℃, water diffusion to SiO_2 becomes significant, and diffusion exponentially increases. Voids at the bonding interface and water molecules in gap can diffuse into the surrounding SiO_2 at high temperatures resulting in a partial vacuum. So wafers will eliminate plastic deformation and eliminate voids. Meanwhile, SiO_2 under this temperature will reduce viscosity, and viscous flow occurs, thus eliminate the micro-gap. When the temperature is above 1 000 ℃, the reaction between mutually adjacent atoms creates covalent bonds, and completes the bond.

Before the bonding stage, surface treatment to make its surface adsorbing is necessary to the silicon wafer. For the thermally oxidized mirror-polished silicon wafer, thermally oxidized SiO_2 has amorphous quartz glass mesh structure. On the surface of SiO_2 film and its inside, there are a number of oxygen atoms in an unstable state. They can obtain energy and leave the silicon atom under certain conditions, and make the surface dangling bonds. There are many ways to increase dangling bonds on the surface of thermal oxidation of silicon. Activation treatment of plasma surface is a method. For the original polished silicon wafer, pure silicon surface is hydrophobic. If it was immersed in a solution containing an oxidizing agent, a layer of single-oxygen-layer will suddenly be adsorbed on the silicon surface. With the increasing of solution temperature (75~110 ℃), the single-oxygen-layer will transfer to monoxide and dioxide. Because surfaces of silicon oxide formed by chemical solution have non-bridging hydroxyl groups, it is good for the wafer bonding at room temperature. Commonly used hydrophilic liquids are sulfuric acid hydrogen peroxide, dilute nitric acid, ammonia etc. For good bonding of silicon, its bonding strength can be up to 12 MPa or more. This requires a good bonding condition. The first factor is temperature. Because bonding of two wafers is ultimately achieved by heating, temperature plays a key role in the bonding process. The second factor is the flatness of the wafer surface. The surfaces of thermal oxidation silicon wafer and polished wafer are not ideal mirrors, but always has a certain surface roughness and undulation. If the wafer has a small roughness, in the bonding process, two pieces will be bonded together completely because of the elastic deformation or viscous reflux effect under a high temperature. If the surface roughness is great, the holes will be produced between the interfaces after the bonding. The third factor is the surface cleaness. If the bonding process is not carried out in a clean environment, there will be some dust particles on the silicon surface. The dust particle is one of the main source of bonded silicon to produce holes. For example, if the wafer thickness is 350 μm, the particle diameter is 1 μm, and the diameter of caused hole is 4.2 μm. From this we can see the dirt particles have a great effect on the bonding result. In addition, gas in the interface will also produce holes at room temperature.

Silicon direct bonding process can not only achieve the bonding of Si—Si、Si—SiO$_2$ and SiO$_2$—SiO$_2$, but also achieve the bonding of Si—quartz、Si—GaAs or InP、Ti—Ti and Ti—SiO$_2$. On the other hand, by mixing an intermediate layer between wafer bonding layers such as low melting point borosilicate glass, we can achieve a low temperature bonding of glass at a low temperature, and the bonding strength is also increased greatly. This low-temperature-bonding technology is compatible with the conventional IC process.

4.5.3 Metal eutectic bonding

The so-called metal eutectic bonding process refers to clipping a layer of metallic material film between a pair of bonding surfaces to form a three-layer structure, then connect to each other at an appropriate temperature and pressure. Commonly used eutectic bonding materials are gold - silicon and aluminium - silicon etc. Figure 4.20 shows four engaging ways of gold - silicon eutectic bonding. Figure 4.20 (a) shows a silicon - gold/(or silicon)-silicon three-layer structure, which can achieve the bonding by Au—Si layer. Figure 4.20 (b) shows the silicon and metal base bonding, which realize the bonding by the eutectic of gold-silicon layer when a layer of gold film is deposited on the substrate. The way of Figure 4.20 (c) is opposite to the way of Figure 4.20 (b). Figure 4.20 (d) shows the firstly made gold-silicon foil (the thickness is 20~40 μm) which is sandwiched between the silicon and the base to achieve the eutectic bonding. It can also use the aluminium - silicon intermediate mezzanine to achieve the three-layer structure eutectic bonding.

Gold silicon eutectic bonding is constantly used in the package of IC technology to wield the material on the substrate. The technique was used on pressure transmitter in 1979. Gold-silicon solder is Au—Si two-phase system (silicon is about 19%), melting point of about 370 ℃, which is much lower than the melting point of pure gold or pure silicon. The eutectic temperature between the aluminium and silicon is close to 600 ℃. In addition to gold, Al, Ti, PtSi and TiSi$_2$ can be also used as a intermediate transition layer of silicon-silicon bonding.

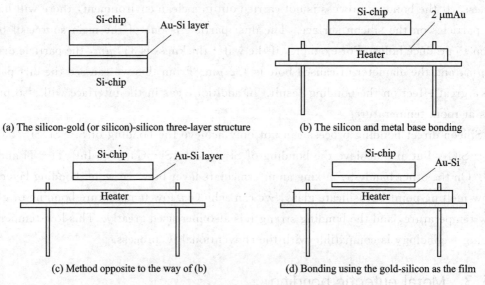

Figure 4.20 Figures of metal eutectic process

4.5.4 Cold pressure welding bonding

The so-called cold pressure welding bonding refers to the method to realize the bonding process by a certain pressure between the elements under the conditions of room temperature and vacuum. Figure 4.21 shows the silicon-silicon cold pressure welding bonding principle.

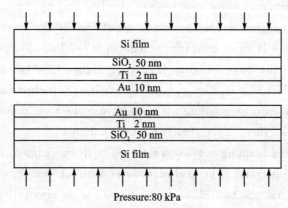

Figure 4.21 The working principle of silicon-silicon cold pressure welding bonding

Before bonding, a layer of silicon dioxide is deposited on the wafer surface, then titanium and gold films are covered on silica. The process is finished under the conditions of room temperature, vacuum and pressure applying state. It can achieve a high mechanical strength.

4.6 Engineering examples of combination for multiple processes to fabricate the MEMS device

4.6.1 Introduction

In actual MEMS device fabrication process, the device generating is a combination of many processes. Generally, the fabrication process can be divided into several parts. Each part is called "layer". The detailed fabrication description with bulk silicon technology considering the present situation of domestic process will be introduced below. This process can be divided into three layers respectively. They are structural layer (which is used to generate a movable device portion), metal layer (which is used to generate a metal wire, then connect the device with peripheral circuits) and the anchor layer. Each part has a certain pattern shape and color as well as a corresponding process (including the photolithography process and the subsequent process: the former is used to pattern transfer from the mask to photoresist; the latter is used to pattern transfer from the photoresist to the substrate material). Each transfer of graphic requires a mask to complete a photolithographic process. This is the basic issue that should be considered.

This subsection focuses on a combination of several processes to introduce the real fabrication process. It includes resonant silicon gyroscope, resonant silicon MEMS pressure sensor and electromagnetic micro-motors etc.

4.6.2 Engineering example of fabrication process for resonant MEMS gyroscope

Working principle of resonant silicon gyroscope will be introduced in Subsection 6.3. In the production process, the first consideration is the research status of the domestic processing.

Because the surface silicon technology is not yet mature in our country, we decide to use bulk silicon process.

In bulk silicon process, taking into account the sensor involving a lot of electrostatic comb drives that have high requirements in size and thickness, including device size, steepness and aspect ratio etc, ICP technology used to ensure the quality of the sensor. In the production of low value and a small-size requirement (approximately 3 microns) device named support chamber, considering the sensor has characteristic of not high production cost and precision of this part, KOH wet etching process is used. At the same time, taking into account the pressure difference between the upper and lower surfaces of mass in ICP technology, four standoffs are designed on the mass (shown in Figure 4.22) to ensure that it will not crush during the process of the fabrication. Meanwhile, the effect of the bonding must be taken into account. Since the bonding process is to combine the two kinds of materials together, the size of the bonding anchor cannot be too small (the minimum bonding size of the design is square and the minimum size is 120 μm $\times$ 120 μm).

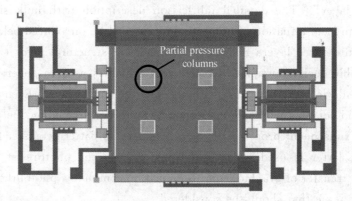

Figure 4.22 Mask sketch of resonant silicon MEMS gyroscope

Taking into account the fabricating level of domestic bulk silicon processing technology, the fabrication process is shown in Figure 4.23.

Firstly, take a 3-inch silicon wafer (shown in Figure 4.23 (a)). Boronizing technology is used to increase its electrical conductivity (shown in Figure 4.23 (b)). After Boronizing is finished, a piece of glass (shown in Figure 4.23 (c)) is selected. By dry etching technology, a shallow groove is made on the surface (shown in Figure 4.23 (d)). On this basis, by sputtering and lift-off technology, the metal electrodes are made on the surface (the material could be $CrAu_4$, as shown in Figure 4.23 (e)).

After the metal electrode is made, the bonding process is used to combine the silicon and glass (shown in Figure 4.23 (f)). By thinning process, the silicon wafer (shown in Figure 4.23 (g)) is thinned. Finally, by ICP process, the final structure is obtained (shown in Figure 4.23 (h)).

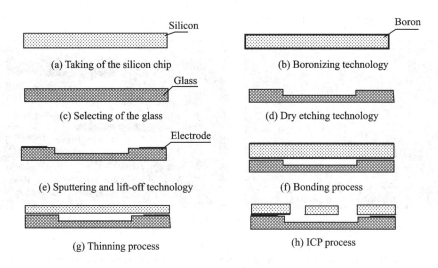

Figure 4.23 Resonant silicon gyroscope processing process

Corresponding to the process, the final mask sketch is shown in Figure 4.22. It is concluded from the sketch that the fabrication process can be divided into three parts (three kinds of colors represent three masks). Firstly, the support cavity must be fabricated to make the subsequently suspended beams (KOH wet etching process). Secondly, the metal wire must be fabricated to achieve the connection with the outside structure. Finally, you need to produce a movable member. Therefore, three processes are needed to achieve the production of the chip. Three processes correspond to three masks. In the process of using L-edit software to make the chip, three masks are represented by three colors, and each color indicates a layer respectively. They are structural layer, metal layer and the anchor layer. Each layer can be individually designed and modified. Three structures can be presented together (it can also be presented separately). Each layer can be modified individually and they do not effect each other.

The generated three masks are shown in Figure 4.24.

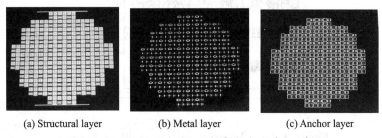

Figure 4.24 Fabricated sketchs of masks

According to this process, the completed sensor micrograph is shown in Figure 4.25.

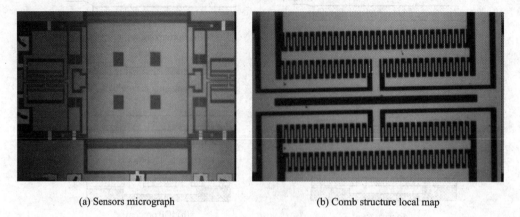

(a) Sensors micrograph   (b) Comb structure local map

Figure 4.25 Resonant silicon micromechanical gyroscope chip architecture

4.6.3 Engineering example of electromagnetic micro-motor production process

4.6.3.1 Construction working principle

Similar to other kinds of electromagnetic motors, planar electromagnetic micro-motor uses the interaction of electromagnetic effects between the stator and the rotor. It relies on the electromagnetic field to change the electrical energy into mechanical energy of the rotor's movement. Figure 4.26 shows the schematic diagram of the motor. As can be seen from the

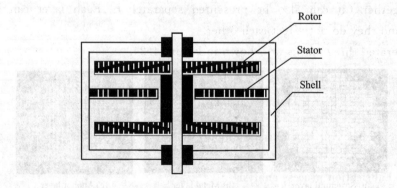

Figure 4.26 Micro-motor assembly schematic

figure, the motor uses a single stator-double rotor brushless DC motor structure. The stator and the rotor are connected in axial direction and fixed on the structure. There are two rotors, symmetrically distributed on both sides of the stator. Armature windings are produced on the stator side. This connection can not only increase the output torque of the motor, but also solve the problem of unilateral axial issue.

Figure 4.27 is the distribution sketch of flat windings. It is concluded from the sketch that the micro-motor stator winding adopts the planar, non-trough, coreless structure concentrated winding. The front and back sides have 18 coils for each side, symmetrically distributed on both sides of the substrate. The connection between the two sides of the coil is through the hole over the line. It works in the same way as a brushless DC motor works.

The stator windings use three-phase star connection (shown in Figure 4.28). Each phase has 12 coils connected in series, according to A, B and C sequentially radially and uniformly distributed on the silicon substrate. The above-described structure of the stator windings drastically reduces the size of the motor in axial direction. Torque ripple caused by the cogging effect for conventional motor is greatly reduced. Stable torque output is also realized. Due to the use of coreless winding structure, hysteresis loss and eddy current losses do not exist. Thereby the transmission efficiency is also improved. Since both ends of the stator coil are directly contacted with the air gap, cooling problem of the coil is also solved. It may have a large electrical load. In order to increase the driving force of the motor, the stator windings are in the form of double-sided type.

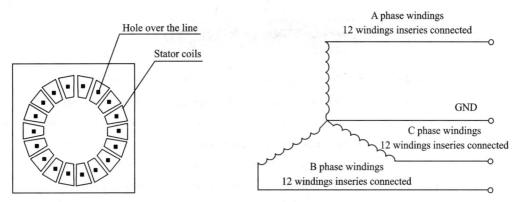

Figure 4.27 Stator coil schematic   **Figure 4.28** Stator windings connection diagram

4.6.3.2 Fabrication process analysis

It is concluded from working principles of the motor that the turns and thickness of stator windings etc. will have a great impact on its output torque. Thus, the stator coils should have a large thickness and a large number of turns. As for the fabrication process of the micro-motor, the main process is the fabrication of a series of double-sided metal stator windings. They are fabricated on double sides of silicon surfaces and each side is fabricated for 18. The two faces are connected by vias. The difficulties of this process are as follows:

(1) How to make high-quality metal coils on double sides of the silicon wafers;

(2) Metal connection between the vias must be guaranteed and the through etching of the monocrystalline must be confirmed.

Therefore, this process has a large difference between the pre-introduced technology of resonant MEMS gyroscope.

Firstly, in the aspects of stator windings fabrication, MEMS sputtering and electroforming process combined method is used. Sputtering process is used to produce the stator coils' seed layer.

Secondly, using a photolithography process, photoresist material is used to fabricate the desired groove which is needed in the electroforming process (shown in Figure 4.29). Because this process requires that the windings have a great thickness to increase that the thickness of the photoresist, the secondary spin coating method is used. It means when the first layer of film is produced, the second layer of film is then produced on it.

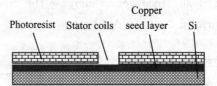

Figure 4.29 Winding electroforming process schematic

Thirdly, the biggest difficulty of the task is the fabrication of stator windings on both sides and good alignment properties on both sides. Thereby two-sided lithography, double-sided alignment methods are used. The technology is simply shown in Figure 4.30.

The fabrication process of the micro-motor consists of two parts. The first part is using Inductively Coupled Plasma (ICP) etching to make via holes of stator windings.

This can be realized by double-sided oxidation, double-sided lithography, development, opening of silica window and ICP deep etching. The second part is using quasi-LIGA fabrication process to make stator windings. This process uses the silicon as substrate. Then, a layer of Cu is sputtered on it to act as the seed layer. By using the double-sided lithography, double-sided development, double-sided electroforming process, 18 windings

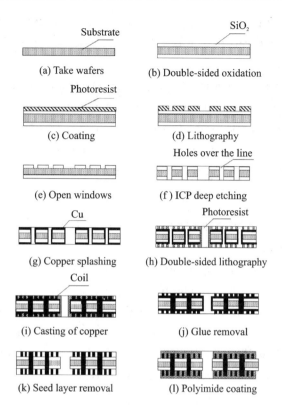

Figure 4.30 Electromagnetic plane micro-motor stator windings production process

are fabricated separately on double sides of the silicon chip. Their thicknesses are about 30 microns. Experimental results indicate that the binding between the stator windings and the substrate is excellent. Conductivity of the windings between positive and negative sides is good. No shedding phenomenon happens for the windings.

It is concluded from the fabrication process that three masks are needed during the course of the fabrication. Three times of lithography are also needed to finish the production. The first mask is used for the production of front windings. The second mask is used to fabricate the windings on the reverse sides. The third mask is used to connect the external test circuit and the windings. Detailed introductions of the processes are as follows.

1. Preparation and pretreatment of silicon wafers

During the production, $<100>$ monocrystalline silicon is selected as the substrate material. Before production, silicon substrate must be clean. The impurities (such as dust, dirt etc.) of the wafer surface should be cleaned. The cleaning can be carried out in a washing machine. After the cleaning, the wafers must be dried to remove moisture on surface. This can help to prevent the loose of photoresist during the course of lithography or development. Drying temperature of the wafer should not be lower than 100 ℃.

2. Silicon oxidation

During the fabrication process of the planar electromagnetic micro-motor, silica is mainly used to act as the mask film for the ICP technology and insulator layer of the stator windings. The thermal growth method is used in the production of SiO_2. The constantly used oxidizing atmosphere is water vapor, dry oxygen and moisture oxygen and so on. Their reactions with silicon under a high temperature are

$$Si + O_2 \xrightarrow{\Delta} SiO_2 \qquad (4.32)$$

$$Si + 2H_2O \xrightarrow{\Delta} SiO_2 + H_2 \uparrow \qquad (4.33)$$

3. Coating technology

Coating technology is an important technology which is used to coat a layer of photoresist film on the silicon chip. The film must possess the characters of good adhesion, suitable and even thickness. If the film is too thin, there will be more pinholes. This will result in the poor ability of corrosion resistance. If it is too thick, the film will result in a low resolution. Karlsuss MA6/BA6 coating machine, which is made in Germany, is used for this process. The coated sketch is shown in Figure 4.30 (c).

4. Lithography

Lithography mainly includes pre-baking, exposure, development and harden etc. Lithography finished graphics window is shown in Figure 4.30 (d).

5. Opening of SiO_2 window

Opening of the silica window in the fabrication of plane electromagnetic micro-motor is mainly used for the preparation of ICP technology. In planar electromagnetic micro-motor production process, hydrofluoric acid buffer etching solution is used. By this means, the photoresist covered part is protected and the uncovered part is etched by the solution. Selection criteria of the silica etching solution is that the etching solution must be able to etch the exposed silicon oxide layer without damage to the photoresist on the surface:

$$SiF_4 + 2HF = H_2[SiF_6] \qquad (4.34)$$

Etching of the silica film with hydrofluoric acid mainly uses complexation reaction between the hydrofluoric acid and silica:

$$Si + 4HF_2 = SiF_4 + 2H_2O$$

$$SiO_2 + 6HF = H_2[SiF_6] + 2H_2O \qquad (4.35)$$

Hydrofluoric acid : Ammonium fluoride : Deionized water = 3 (Milliliter) : 6 (Gram) : 10

(Milliliter).

The opened silica window is shown in Figure 4.30 (e).

6. Fabrication of the line hole by ICP process

Production of electromagnetic plane micro-motor over the line hole can use wet etching (such as KOH corrosion) or dry etching (such as ICP) technology. Although KOH wet etching has advantages of easy to operate and high corrosion rate, due to the silicon crystal orientation effect during etching process, verticality of the line hole is difficult to ensure. In addition, due to the presence of lateral corrosion and characteristics affected by temperature, it is hard to be controlled. The dry etching may well overcome these shortcomings. Thus the dry etching (such as ICP process) is used to fabricate the line hole. The fabricated ideal hole is shown in Figure 4.30 (f).

7. Copper splashing

In electromagnetic planar micro-motor production process, copper splashing technology is mainly used to fabricate a layer of copper-seed-layer for the electroforming process. This process is carried out in SP-I type 2 kW RF matching unit. The finished structure is shown in Figure 4.30 (g).

8. Double-sided lithography

According to the structure and working principles of plane electromagnetic micro-motor, the stator windings are symmetrically distributed on two sides of a silicon substrate. The stator windings must be fabricated on two planes and the corresponding windings between the two sides must be aligned. The solution is to make two cross marks on double sides of silicon chip. Then, by using the performance of double lithography of the machine, this will be realized easily. The completed two-sided winding schematic is shown in Figure 4.30 (h).

9. Copper electroforming

Electroforming process must be carried to obtain the required stator windings. This process is carried out in self-developed electroforming instrument. Copper sulfate solution is used as the electroforming solution. Through trial and error, come to the ideal conditions as Figure 4.30 (i) shows. The fabricated final stator windings are shown in Figure 4.30 (j).

10. Glue removal

After the electroforming of stator windings is completed, the photoresist on surface must be removed to get rid of the seed layer and then realize the insulation between the wires. This can be realized by putting the wafer into the acetone solution and then clean it with water or alcohol. The fabricated plane stator coil can be obtained then.

11. Seed layer removal

Seed layer removal process is used to achieve the insulation between the stator windings. In planar electromagnetic micro-motor production process $FeCl_3$ solution is used to remove the seed layer. The reaction can be expressed as

$$FeCl_3 + Cu \rightarrow FeCl_2 + CuCl_2$$

In the process of seed layer removal, Ferric chloride solution reacts with the copper in the seed layer and the stator coil at the same time. However, due to the thickness of the seed layer (about 2 000 Angstroms) is negligible comparing to stator windings (approximately 30 microns in thickness), there is almost no loss for the stator windings when the seed layer is removed. To prevent to react too fast, ferric chloride is generally used in low concentrations (less than 10%). The completed schematic of seed layer is shown in Figure 4.30 (k).

12. Polyimide coating

After seed layer of flat stator coil is removed, 18 planar coils on both sides of the silicon substrate are generated. They must be protected to avoid the friction or breaking of the wire. The protection can be achieved by coating a polyimide insulating material layer on coil, the polyimide film coated stator coil is shown in Figure 4.31.

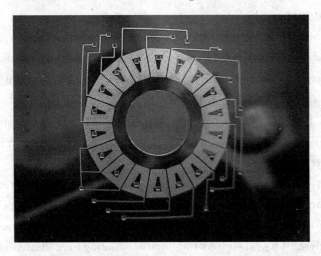

Figure 4.31 The photo of 18-stator-coil windings

4.7 Summary

In this chapter, the fabrication process of MEMS is introduced in detail. Many engineering examples are introduced combing the author's working experience. Detailed processes include the lithography process, bulk silicon process, surface silicon process, LIGA process, sputtering technology, lift-off process, film preparation process and bonding technology. At the end, detailed engineering examples of MEMS gyroscope and MEMS electromagnetic micro-motor are introduced.

References

[1] Deal B E, Grove A S. General Relationship for The Thermal Oxidation of Silicon[J]. Journal of Applied Physics,1965,36(12):3770-3778.

[2] Campbell Stephen A. The Science and Engineering of Microelectronic Fabrication[M]. Oxford: Oxford University Press, 2001.

[3] Liu Zhong'an. Optimization of Photolithography Process Parameters[J]. Semiconductor Optoelectronics,2001,22(1):52-54.

[4] Wang Yangyuan, Kang Jinfeng. Development and Challenge for The Lithography Technology of Silicon Integration Circuits[J]. Chinese Journal of Semiconductors, 2002 (3):225-237.

[5] Chen Dongshi. Guarantee for Lithographic Accuracy of Processing Superfine Line[J]. MICROPROCESSORS,2002(2):23-26.

[6] Electronic Industry Semi-conductor Technology Book Group. Semi-conductor Element Technology[M]. Shanghai: Shanghai Technology Literature Press, 1984.

Chapter 5
Friction wear and tear under micro scale

MEMS (Micro-Electro Mechanical System)[1-3] is a new research field developed in recent decades. It has great prospect of applications in many fields such as industry[4-5], information and communication[6], national defense, aeronautics and astronautics[7-8], navigation, medical treatment[9-10], bio-engineering, agriculture, environment and domestic service etc. It has attracted the worldwide attention and many academic institutions are carrying on the pertinent researches. MEMS has become another major high-tech industry after microelectronics technology.

However, with the dimension decrease of the MEMS device, the ratio of surface area to volume increases. Then, unlike macro conditions, its mechanical characteristic displays strong size effect[13-14], which makes surface effects greatly enhanced. For example, when the dimension of structure decreases from 1 mm to 1 μm, its area decreasing factor is 1 million times, and its volume decreasing factor is 1 billion times. Thus, compared with the forces such as the inertial force and the electromagnetic force, which are proportional to the device's volume, frictional force and adhesive force that are proportional to its area increase several thousands times and become the main forces to MEMS. So, large-scale systems are influenced by inertia effects to a much greater extent than small-scale systems, while small-scale systems are more influenced by surface effects. As for the movable MEMS devices[15-16], because the gaps of the contact pairs are in nanometer scale even zero, influence of the frictional force between the frictional pair increases greatly. This not only limits the dynamic performances of the micro device, but also aggravates the surface injury of the contact pair. As a result, the working performance, the operation life and the reliability of the movable MEMS device are seriously hampered. It is necessary to carry on the micro-tribology research in order to avoid this phenomenon. On the other hand, because the conventional frictional law is no longer feasible to micro-structure, it is vital to rebuild the micro-tribology theory in which the size effect must be concerned. Correspondingly, an applicable experimental method which can successfully obtain the credible data for micro-tribology research is also of great importance.

Generally, in point of the integration degree, there are two kinds of methods to measure the tribology characters of the movable MEMS devices. One kind is on-chip measuring method and the other is off-chip measuring method. This chapter introduces the working principle of each method and comments many testing results for each method.

5.1 Off-chip testing method for micro friction

Off-chip testing method refers to the method that the micro specimens to be tested are fabricated with MEMS technology and the experiment is carried out with the external testing devices. In this method, the installation and clamp of the specimens are performed by the device installed on the equipment and the tribology data is obtained by the force sensor fixed on it. Generally, there are three kinds of off-chip measuring methods—the pin-on-disc method, the AFM (Atomic Force Microscopy) testing method and the special testing method etc.

5.1.1 Micro-tribology test with the pin-on-disc measuring method

The so-called pin-on-disc measuring method is one of the methods which are constantly used in tribology experiments. Figure 5.1 shows a typical pin-on-disk micro-tribology testing device fabricated at Institute for Technology, Hannover, Germany. This device is mainly composed of a rotary stage to place the samples, a motor to drive the rotary stage and the force testing system. In order to measure the frictional force between the frictional pair, the pin-on-disk tester is equipped with a commercial optical frictional force measurement sensor (shown in Figure 5.1 (b)). It consists of a spring system allowing a relative deflection to measure the load and a fiberoptic system to measure the degree of deflection. During the course of the experiment, one surface of the contact pair is fixed on the rotary stage and another is fixed on the test head (shown in Figure 5.1 (b)). Normal force between the contact pair equals to the restoring force caused by the deformation of the suspended beam in vertical direction. Its magnitude can be regulated by changing the displacement result in vertical direction. Then, rotate the workbench slowly to make a comparative rotation between the contact pair. The frictional force can be obtained by measuring the horizontal deformation of the suspended beam by the optical testing sensor. Hans H. Gatzen et al.[17] measured the tribology characters between the surface of hard disk and the magnetic head with this method. During the course of the experiment, the material of the disk is aluminium chip coated with the films customary on rigid disk media. The two top layers are Diamond Like Carbon (DLC) films coated with a lubricant. The sliders fixed on the test head are two single crystal silicon chips with different surface topographies. One of them is machined flat by means of nanogrinding and the other is fabricated with three round spots to simulate the three-point contact. Magnitude of the normal force is in the range of $10 \sim 60$ mN. It is concluded from the tested result that the frictional coefficient between the disk and the silicon chip with unpatterned surface, which is 0.24, is smaller than that between the disk and the patterned silicon chip, which is 0.34. The frictional coefficient is the function of the normal load applied on the frictional pair. The increase of the apparent contact area results in a decrease of friction.

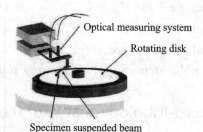

(a) Sketch of the equipment   (b) Enlarged portion of the testing portion

Figure 5.1   The testing equipment of the pin-on-disc measuring method

Though this method can be used to measure the tribology characters of some movable MEMS devices, there is one shortcoming of it. Because the measuring scope of the force is in mN, it is hard to precisely measure the frictional force of MEMS devices whose frictional force is in micro Newton order. Therefore, this method is mainly used to measure the tribology characters of the macro devices.

5.1.2   Micro-tribology test with AFM

Atomic Force Microscopy (AFM) is one of the high precision commercial off-chip measuring equipments used to measure the tribology characters of movable MEMS devices at present. This device (its working principle is shown in Figure 5.2) uses the weak interaction (van der Waals force) between the tip and the surface of the specimen to be measured. During the course of the experiment, the sample to be tested is placed on the sample platform, and the contact pair is composed of the top surface of the sample and the tip of AFM. The tip of AFM is fixed on V-shaped micro-cantilever which is very sensitive to the micro force. The other end of the cantilever is fixed on the structure. Firstly, adjust the position of the sample table to make the tip of AFM touch the surface of the specimen slightly. Then, slide the tip on the surface of the sample slightly. During the course of its sliding top end of the tip will initiate a little displacement in horizontal and vertical directions and their magnitudes can be measured by the optical measuring system. The optical system transforms the optical signals into electric signals. Then the enlarged electric signals are transformed into the frictional force and normal force by the control circuit. Their tribology characters can also be obtained then. Many institutions are carrying on the tribology research with this method and they obtained many useful conclusions. In 1996, Bharat Bhushan et al.[18] in Ohio State University in the United States compared the tribology characters of the pure single crystal silicon with different crystalline orientations (Si (111), Si (110) and Si (100)), the PECVD-oxide silicon samples, dry-oxidized, wet oxidized and ion-implanted silicon samples. It is concluded from the tested result that the PECVD-oxidized samples have the largest wear resistance followed by dry-oxidized, wet-oxidized, and ion-implanted samples. The crystalline orientation of silicon has little influence on the frictional coefficient of the single crystal silicon. The frictional coefficient, which is 0.04, is much lower than that tested in macro condition (about 0.3). The explanation for this is in micro condition, the material's real contacting area reduces, and then the number of contacting points reduces correspondingly. This leads to the influence of ploughing effect to frictional force greatly weakened. They also found that with the increment of the normal force applied on the contact pair, the frictional coefficient gradually close to the macro condition while the damage of the contacting surface greatly aggravated. This conclusion is contrary to the conventional Amontons law and further shows that it is necessary to research the tribology characters of MEMS devices. In 2000[19], the research team discussed the influence of surface topography to the micro frictional coefficient. They found the surface fabricated with a certain topography has greater influence to the frictional coefficient than the un-fabricated

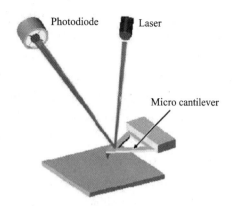

Figure 5.2 Working principle of the atomic force microscope

surface. Bhushan explained this with the so-called ratchet mechanism of friction. In his opinion, the difference of the frictional coefficient is mainly caused by the difference of the contacting angle between the tip and the surface when the tip moves back and forth along the contacting surface. When the gradient of the asperity is positive during the course of the reciprocating motion of the tip, their frictional coefficient will correspondingly increase. On the contrary, when the asperity has a negative gradient, the friction coefficient will decrease. This explanation refutes the former standpoint which thought the influence of surface topography on friction coefficient had nothing to do with the scan direction of the tip.

Though there is a large progress in the study of the friction and wear for micro device with AFM, there are some unsolved problems for this method. Firstly, because the contact condition of the tip and surface is point-surface contact, this can not successfully simulate the real contacting condition of MEMS devices. Secondly, because the real contacting surfaces of the MEMS devices are fabricated with MEMS technology, it is not totally correct to think the measured data can successfully represent the real data of MEMS device. Thirdly, the stiffness of the tip's cantilever may influence the results.

5.1.3 Micro-tribology test with special measuring device

Besides the pin-on-disc experimental method and the AFM method, there is another method—the private tribology measuring method. This method uses the micro-friction testing devices developed by the institutions themselves. Apart from the former two kinds of measuring methods, the contacting surfaces in this method are fabricated with MEMS technology to successfully simulate contacting condition of MEMS devices. Clamping of the specimens, pre-loading of the contact pair, picking of the tribology data (obtained from the force sensor installed on the equipment) are all carried on by this equipment. There are many research institutions such as IMT(Institute for Micro Technology) in Germany, Hannover University and University of California etc.[20-23] carrying on the experiment with this method and great progress has been achieved. Chen Quanfang[23] et al. in University of California tested the tribology characters of single crystal silicon with this method. Simple sketch of this device is shown in Figure 5.3. In this method, one surface of the contact pair is the unprocessed single crystal silicon while the other is the single crystal silicon chip whose top surface is fabricated with a series of columns by MEMS technology. The columns are 3 μm in height and the diameter varies from 2 μm to 1 024 μm. The tested relationship between the contact scale and the frictional coefficient is shown in Figure 5.4.

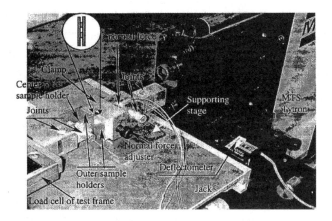

Figure 5.3　Setup of the special micro force test system

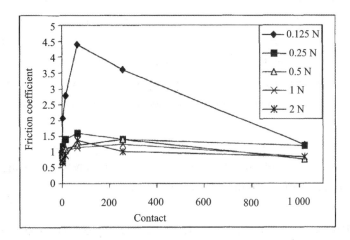

Figure 5.4　Relationship between the contact area and the frictional coefficient

It is concluded from the sketch that when the diameter of the column is less than 64 μm, the frictional coefficient increases with the increment of the diameter and reaches its maximum value when the diameter reaches 64 μm. Then, with the increment of the diameter, the frictional coefficient gradually decreases with it until the diameter reaches 256 μm. Then, the frictional coefficient tends to be a constant when the scale surpasses 256 μm. This means when the diameter is smaller than 256 μm, the conventional frictional law is no longer feasible then. They also found the crystallographic direction of single crystal silicon had great influence on the frictional coefficient (surface of the (111) single crystal silicon possesses higher frictional coefficient than the (100) chip (10%～60%)).

The advantage of this method is that it can successfully simulate the real contact condition of MEMS devices. However, dimension of the specimen is so little (in micrometer order) that it is hard to precisely fix the frictional pair. This may result in some measuring error for the tested result. At the same time, this method mainly imposes on the top surface of the MEMS device and it is hard to simulate the sidewall contact condition of the movable MEMS devices.

5.2 On-chip testing method for micro friction

On-chip testing method refers to the method that the loading apparatus, testing apparatus and force sensors are all integrated on a single chip with MEMS technology and the tribology experiments are performed by the system itself. Comparing with the off-chip measuring method, the on-chip measuring method has the advantages of easy fabrication, high resolution of micro force and displacement etc. At the same time, because the contacting surfaces composing the frictional pair are all fabricated with MEMS technology, the contacting situation can successfully simulate the real contact condition of the MEMS device. Thus the on-chip measuring method has become one of the most important measuring methods for the micro tribology research.

Generally, according to the driving method of dynamic source, the on-chip measuring method can mainly be classified into the electrostatic driving method, using the difference of the mechanism characters of the bimorph material etc.

5.2.1 On-chip testing method actuated by electrostatic force

This method mainly uses the electrostatic force as the driving source. Corresponding structure to initiate the electrostatic force can be the plate capacitance or the comb actuator. According to the contacting condition of the contact pair, this method can be classified as the plane-plane contact, plane-cylinder contact, line-line contact and plane-sphere contact etc. Difference of the contact pair may lead to the difference of the tribology character.

5.2.1.1 Plane-plane contact testing method

Plane-plane testing method mainly refers to the method that the two contacting surfaces are planes during the course of the experiment[24-27]. Figure 5.5 is one of the typical Scanning Electron Microscopy (SEM) sketches of the plane-plane contact testing method designed and fabricated by Tas N. R. et al. of Twente University, the Netherlands[24] (enlarged portion of

A is shown in Figure 5.5(b)). This device is mainly used to prove the correctness of their friction model. Material of this micro-tribosensor is polysilicon. In this figure, the driving sources are two perpendicularly placed comb finger actuators C and D. During the course of the experiment a DC driving voltage is firstly applied on actuator D to make the shoe-shaped contacting head B contact the rectangle static block under the function of the electrostatic force. Value of the normal force equals to the restoring force of the suspended beam caused by its relative displacement. Then, gradually increasing the DC driving voltage applied on comb actuator C, the frictional force can be obtained when there is a relative displacement between the contact pair. The frictional coefficient can also be obtained then. This method has the common advantages of the on-chip testing apparatuses such as high degree of integration, easy fabrication etc. However, because the comb actuator in this sensor applies the single cantilever structure, stability of this system can not be confirmed for the comb teeth are easy to contact. This may result in the performance failure of the whole system. This method must be improved in some aspects.

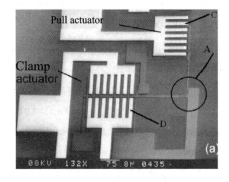
(a) Total sketch of the micro-friction tester

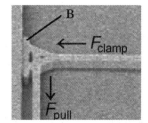
(b) Enlarge portion of the contact pair

Figure 5.5 SEM sketch of the surface-surface contact type on-chip micro-friction tester

5.2.1.2 Plane-cylinder contact testing method

Plane-cylinder testing method refers to the method that one surface in the contacting pair is plane and the other surface is semi-cylinder type. Similar to the plane-plane contacting method, this method focuses on the tribology characters research of sidewall contacting pair for MEMS devices. Figure 5.6 is the SEM sketch of the plane-cylinder contacting method (enlarged portion of the contacting pair is shown in Figure 5.6(b)). The testing device, whose material is polysilicon, is fabricated in American Sandia National Laboratory[28-31].

This method is similar to the plane-plane method fabricated at Twente University, but it uses the symmetrical cantilever to support the whole system and the problem of instability is avoided.

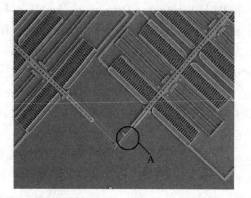
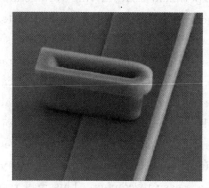

(a) Total sketch of the micro-friction tester   (b) Enlarge portion of A

Figure 5.6 SEM sketch of the planar surface- cylindrical surface contact type on-chip micro-friction tester

Meanwhile, as the fixed semi-circular column anchor is designed at the inner side of the extended arm of the comb finger actuator, the normal force must be applied by pulling the extended arm of shuttle, but not by pressing. The measuring error caused by the pressing way is also avoided then. At the same time, because the comb fingers in the system move along the direction of the fingers of the shuttle, but not perpendicularly to them, displacement of the moving part is greatly increased then. This confirms the excellent testing result of the system. Donna Cowell Senft and Dugger Michael T. et al.[28] in American Sandia National Laboratory studied the difference of the tribology characters between the pure silicon-silicon contacting pair and the pair which is fabricated with ODTS (octadecyltrichlorosilane) on the surfaces. It is verified from the tested results that the life span of the ODTS fabricated surface is three times the length of the uncoated surface while their frictional coefficient, 0.14~0.16, are very close to each other. After that, they coated another kind of material on the contacting pair PFTS (perfluorodecyltrichlorosilane, $C_6F_{13}CH_2SiCl_3$)[29]. From the tested result they found in the dry air condition, the stable frictional coefficient is 0.06, which is much lower than the ODTS coated surface (shown in Figure 5.7). But, in the water vapor filled condition, the friction behavior of the device was erratic, with the frictional coefficients between zero and 0.3 observed. Failure happened after 10^5 cycles' run. Accumulation of the wear material (shown in Figure 5.8) can be observed by SEM. No wear

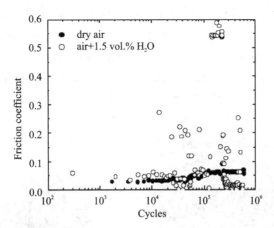

Figure 5.7 Friction coefficient versus cycles for PFTS coated sidewall friction testers in dry air and humid air

accumulation was observed in dry air conditions. In 2006, Guo Zhanshe, Meng Yonggang et al.[32] in State Key Laboratory of Tribology, Tsinghua University, China, studied the tribology characters of single crystal silicon MEMS devices with a special tribo-sensor. SEM sketch of the device is shown in Figure 5.9. In this device two surfaces composing the contact pair are fixed on the extended arms of two comb finger actuators respectively. One comb actuator in the

(a) Wear surfaces on the beam in dry air   (b) Wear surfaces on the post in dry air

(c) Wear surfaces on the beam in humid air   (d) Wear surfaces on the post in humid air

Figure 5.8 Scanning electron microscope images showing the wear surfaces

system offers the normal force by moving the structure in vertical direction (enlarged portion of the contact pair is shown in Figure 5.9(b)). Then, the other comb finger moved back and forth to initiate the frictional force of the contact pair. From this device they found the frictional coefficient between the pure single crystal silicon is about 0.9 and the dynamic frictional coefficient changes with the variation of normal force, which is contrary to the conventional tribology law. They explained this phenomenon with "size effect".

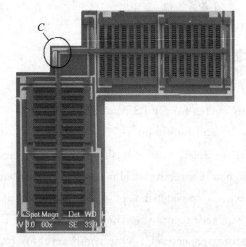
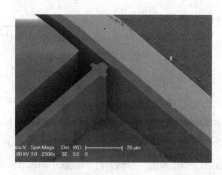

(a) Total sketch of the single crystal silicon tribo-sensor   (b) Enlarged portion of the contact pair

Figure 5.9 Scanning electron microscope image of the single crystal silicon tribo-sensor

5.2.1.3 Plane-sphere contact testing method

In this method, the contact surfaces formed the contact pair are a plane and a hemisphere surface. Similar to the former introduced on-chip measuring methods, initiating of the frictional force in this method is also caused by the movement of the comb finger actuator. The different point is that the normal force (also the electrostatic force) between the contact pair is obtained from attraction between the parallel placed plate capacitors, while in the former introduced on-chip measuring methods, the normal force is initiated by the comb actuator. Figure 5.10 is the SEM sketch of one of the typical measuring structures fabricated by Lim M. G. et al.[33-36] from University of California at Berkeley in the United States. This microstructure, which is made of polysilicon, is fabricated with the sacrificial layer technology of surface process. The comb actuator in the system applies the symmetrical suspended beam to increase the stability of the system. At the bottom of the extended arm of the comb actuator, four semi-sphere shaped dimples (side view of the dimples is shown in

Figure 5.10 (b)) are symmetrically distributed against the four plates which are fabricated under them to initiate four contact pairs. In the middle of one end of the comb actuator, there is a plate capacitor. During the course of the experiment a DC driving voltage is firstly applied on the comb actuator. Then, under the action of the electrostatic force, the moving part of the actuator initiates a relative displacement. Magnitude of the electrostatic force can be calculated by the deformation of the suspended beam. Then, a DC driving voltage is applied on the electrode of the substrate (the value of the driving voltage must confirm the static frictional force is caused by the normal force is larger than the restoring force caused by the displacement of the suspended beam) to make the substrate contact with the semi-sphere (the magnitude of the normal force changes with the driving voltage). Then, remove the DC voltage imposed on the comb actuator. Subsequently, gradually decrease the driving voltage applied on the substrate to reduce the normal force is applied on the contact pair. Record the value of the driving voltage as soon as relative displacement happens between the contact pair. Normal force at this moment can be calculated by the capacitance theory and the corresponding frictional force can be obtained by the deformation of the suspended beam. The research team tested the static frictional coefficients on the condition of dry air and vacuum respectively. It is verified from the tested result that the former tested value (0.58), and much bigger than that of the latter result (0.43). They thought this happened because the surface force affected by the humidity was larger in the air than in the vacuum.

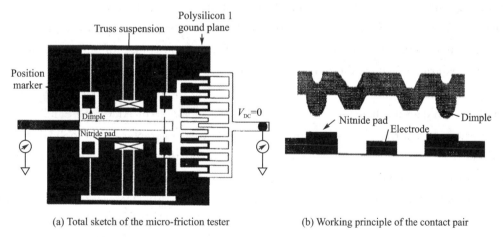

(a) Total sketch of the micro-friction tester (b) Working principle of the contact pair

Figure 5.10 Sketch of the planar surface-spheric surface contact type micro-friction tester

This kind of method can successfully measure the static frictional coefficient of the contact pair. But, the device can not measure the dynamic frictional coefficient of the MEMS devices because of the limit of the fabricated structure.

5.2.1.4 Line-line contact testing method

Line-line contact testing method mainly refers to the method which is used to measure the micro-wear characters of the micro-gear or the micro rotary axis in MEMS. Because the frictional pair in this method is linear contact, we call them line-line contact testing method. Figure 5.11 is the SEM sketch of the micro-sensor fabricated by Tanner Danelle M. et al. in

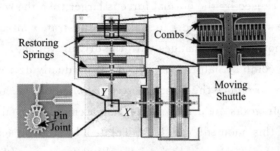

Figure 5.11 SEM sketch of the line-line contact testing method fabricated at Sandia National Laboratory

American Sandia National Laboratory[37-42]. The material of this sensor is polysilicon. Like the plane-cylinder method, this component also uses two perpendicularly placed comb actuators as the power source. Extended arms of the comb actuators are perpendicularly connected. At the tip of the extended arm of the vertically placed comb actuator, a hinge configuration structure is used to link the gear. During the course of the experiment, the square waves are applied on the two comb actuators according to a certain order. Then, the linear movement of the comb actuators is exchanged into the rotating speed of the micro-gear through the linkage mechanism. Tanner Danelle M. et al.[40] in American Sandia National Laboratories studied the effect of air humidity to the wear character of the polysilicon based MEMS devices using this method. From the tested result they find that a very low humidity can lead to very high wear without a significant change in reliability. Volume of wear debris is the function of the environment humidity. As the humidity decreases, the wear debris increases with it. They thought this happened because in the higher humidity conditions, the surface hydroxides in the water vapor might act as a lubricant. The dominant failure mechanism for this MEMS device has been identified as wear. Wear debris has been identified as amorphous oxidized silicon. The constitution of the wear debris is shown in Figure 5.12. It is tested by using a Phillips CM30 300 keV transmission electron

Chapter 5 Friction wear and tear under micro scale

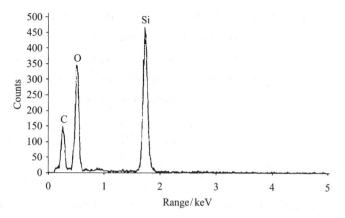

Figure 5.12 EDS spectra of wear debris found outside gear (TEM) with an attached energy dispersive x-ray spectroscopy (EDS) system microscopy.

This method is only used to measure the wear characters of the MEMS devices. For the frictional test, this method is not feasible because of the structure restrict.

5.2.2 On-chip micro-friction testing method using the mechanism characters of the bimorph material

Besides the electrostatic comb actuator driving method, there is another kind of measuring method which uses the difference of the mechanism character of the bimorph material. Figure 5.13[43-45] is one of the typical devices fabricated by using the Multi-User MEMS Process (MUMPs) at the Microelectronic Center of North Carolina (MCNC). This

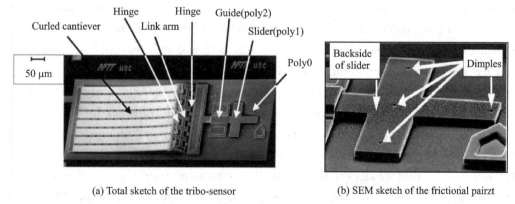

(a) Total sketch of the tribo-sensor (b) SEM sketch of the frictional pairzt

Figure 5.13 One typical device fabricated by using MUMPs at MCNC

structure is mainly actuated by a lateral micro-motor. In this micro-motor, a large curled cantilever serves as the fundamental drive mechanism. The cantilever is composed of the gold-polysilicon bimetallic material fabricated by the sacrificial layer process of the surface technology. This may result in the high precision sensitivity to the temperature because of the difference of the mechanism character. A link arm, which is attached to the tip of the cantilever and the output slider by the hinges, converts the vertical deflection of the cantilever into the lateral motion of the micro-motor. Because the slider supporting structure is very long, a series of dimples are fabricated under it to decrease the effect of the mass. The frictional pair consists of the triangle static anchor and side surface of the slider (shown in Figure 5.13 (b)). The electrostatic cantilever motor is driven by applying voltage across the cantilever and underlying conductive substrate. Patton T. et al.[43] studied the failure mechanism of this device at the various relative humidity (RH). They found that both the electrical and tribological performances of the motor depended on the RH. Electrostatic force acting on the capacitive device is found to decrease with the increasing RH. In a dry 0.1% RH environment, excessive wear of sliding contact and high adhesion caused early failure. In moderate environments up to 50% RH, negligible wear and low adhesion are observed, and the durability was 3 orders of magnitude larger than in the dry environment. The motor rotates 10^5 cycles at 0.1% RH and 10^9 cycles at 50% RH. At 70% RH, the motor can not work because of the high sticking of the device. They believed this happened because the water vapor in this experiment behaved as a gas phase replenishable lubricant by providing a protective adsorbed film. This also indicates that the environment selection of RH is very important to the life-span of the micro-motor. Later, they fabricated a layer of octadecyltrichlorosilane (OTS) self-assembled monolayer (SAM) coating on the lateral surface of the frictional pair to change surface chemistry[44]. By comparing the tribological performance of uncoated and OTS-coated condition, they find that the OTS coating broadened the operating envelope to 30%~50% RH and reduced stiction, which allows better dynamic performance at high RH. The OTS coating improved durability at 0.1% RH is still poor. At high RH, stiction problems reoccurred when the OTS coating is worn away. By controlling and balancing surface chemistry (adsorbed water and OTS), excellent performance, low friction and wear, and excellent durability are attained with the lateral output motor. In 2005 or so[45], they successfully coated a layer of diamond like carbon (DLC) coating to protect against the wear. It is concluded from the tested result that the DLC coating maintained low friction is longer than the uncoated silicon. DLC on DLC experiment shows the lowest friction, and those running in 30% RH show a much longer lifespan than the ones running in dry air. A very clear difference in wear debris is seen between devices running in air and in vacuum. Cylindrical rolls are dominant in the devices that are running in air while platelets are dominant on devices running in vacuum. Ultimately, the DLC coating is found greatly improved the performance of MEMS device.

5.3 Example of the design for an on-chip micro-friction structure

This subsection introduces a kind of silicon based on micro-friction measurement system. This design takes the electrostatic comb finger actuator as the original driving power. One of them acts as the loading unit and the other acts as the friction force initiating structure by back and forth movement. This subsection introduces its working principle and calculation process in detail. On the other hand, in order to confirm the reliability of the design, FEM method is used to make the simulation.

5.3.1 Structure and working principle

Simple working principle of the system is shown in Figure 5.14. This device is mainly composed of two perpendicularly placed comb finger actuators: actuator A and actuator B. Every actuator takes the middle beam as the axial symmetrical line. Actuator B acts as the loading unit to offer the normal pressure force which is needed for the frictional measurement. Actuator A acts as the driving unit to offer the back and forth movement for the measurement. Every comb finger actuator is supported by six supporting beams. Supporting devices connect the substrate by the anchors.

The symmetrically placed beam confirms the driver moving along the axis. It can act as the direction leading performance. Extended portion of the actuators is composed of a friction pair (shown in Figure 5.14 (b)). Extended portion of actuator B is a "+" type semi-cylinder structure. Because there is a 3 μm gap between the semi-cylinder surface and the extended portion of actuator A, we call it the separating type micro-friction measurement system. In order to conveniently solve the tested data, a fixed rectangle block D is set near the cross position. During the course of the measurement a DC voltage is firstly applied on actuator B to make it tightly pressed on the surface of extended portion of actuator A. Thus the contact pair is formed. The contact force can be calculated by the relative displacement of the extended portion of actuator A comparing to D. Then, by applying a DC or sine voltage on A, the relative back and forth movement can be carried between the lateral side of A and

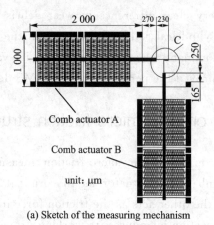 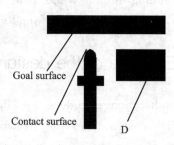

(a) Sketch of the measuring mechanism   (b) Enlarged portion of testing part

Figure 5.14 Working principle of the separating micro-friction testing system

extended portion of B. Friction force can be calculated by the relative displacement. Ration between the normal force and the friction force can be calculated correspondingly. This system integrates the measuring element, the loading mechanism, the driving mechanism and the force sensor on a single chip. It can not only reflect the real contact condition of MEMS devices, but also conveniently obtain the testing data.

5.3.2 Calculation of pertinent theory

Calculation of the silicon based on micro friction measuring system includes the structural stiffness, resonant of the actuator, critical driving voltage calculation, friction force calculation and normal force calculation etc.

5.3.2.1 Stiffness calculation of the comb actuator and simulation

1. Stiffness of the comb actuator

Comb finger actuator is the force source and the key constituent of the system. Mechanical character such as the stiffness etc can initiate a great effect to the measuring result. It is necessary to make the stiffness calculation to a single actuator. Corresponding calculation model is shown in Figure 5.15. Because the three pairs of supporting beams are the same, we can select one pair for the calculation. Total stiffness is three times of one pair.

Simplified model of the calculation model is shown in Figure 5.16. Length of the supporting beams can be taken as l and the applied force is F. The decoupled force sketch of point C is shown in Figure 5.17. From the sketch we can see at point C, the decoupled forces are F_x in

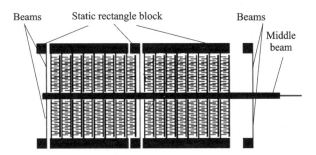

Figure 5.15 Calculating model of the stiffness for the comb actuator

x direction, F_y in y direction and moment of couple M_C. Boundary condition is that when F is applied, the displacement on x and y directions is zero to point C. Using the Moore

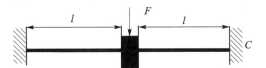

Figure 5.16 Simplified calculation model of the beam

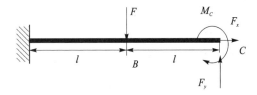

Figure 5.17 Decoupled model

integral method, when the unit force perpendicularly downward is applied on C, the displacement on point C can be expressed as

$$\delta_y = \frac{1}{EI}\int_0^l (F_y \cdot x - M_C)x\,dx +$$

$$\frac{1}{EI}\int_0^l [F_y l - M_C + (F_y - F)x](l+x)\,dx = 0 \qquad (5.1)$$

where E, I and l are elastic modulus of silicon, inertial moment of the beam along the bending direction and the beam length respectively. Equation (5.1) can be simplified as

$$16F_y l = 12M_C + 5Fl \qquad (5.2)$$

Using the same method, when F is applied on C and taking zero as the boundary condition for the rotating angle, equation of the rotating angle can be calculated using the Moore integral method at C:

$$\theta_C = \frac{1}{EI}\int_0^l (F_y \cdot x - M_C)\,dx +$$

$$\frac{1}{EI}\int_0^l [F_y l - M_C + (F_y - F)x]\,dx = 0 \qquad (5.3)$$

Simplified equation (5.3) can be expressed as

$$4F_y l = 4M_C + Fl \qquad (5.4)$$

Combining equations (5.2) and (5.4), the relationship among F_y, M_C and F can be expressed as

$$F_y = \frac{F}{2} \tag{5.5}$$

$$M_C = \frac{Fl}{4} \tag{5.6}$$

Correspondingly, when the unit force is applied on B, displacement caused by F on this point can be expressed as

$$\delta_B = -\frac{1}{EI} \int_0^l [F_y l - M_C + (F_y - N)x] \cdot x \, \mathrm{d}x = \frac{Fl^3}{24EI} \tag{5.7}$$

Thus, the stiffness on the moving direction can be expressed as

$$k_s = \frac{F}{\delta_B} = \frac{24EI}{l^3} \tag{5.8}$$

Because there are three pairs of supporting beams on the comb actuator, the total stiffness can be expressed as

$$k = 3k_s = \frac{3F}{\delta_B} = \frac{72EI}{l^3} \tag{5.9}$$

Thus, the calculated stiffness of the supporting beam is 71 μN/μm.

2. Calculation of critical driving voltage

Critical force refers to the minimum driving voltage applied on comb actuator B in order to make the friction pair contact to each other. This parameter is one of the most important indexes to check the performance of the system. Value of the critical voltage is pertinent to the designed original gap of the contact pair. The less the original gap is, the better the measuring performance will be. But, if the gap is too small during the course of the design, the fabrication will be difficult and the cost will increase correspondingly. If the gap is too big, the critical voltage will be too large and the breakdown will happen on the silicon chip. The actuator will be failure then. So, it is important to deduce the relationship between the gap and the critical driving voltage.

According to the plate capacitance theory, when the edge effect is ignored and a driving voltage is applied on the comb actuator, the relationship between the static force F_{static} and v can be expressed as[46]

$$F_{\text{static}} = \frac{n \varepsilon t v^2}{g} \tag{5.10}$$

where n is the number of pairs, ε is the permittivity constant of air, t is the height of the comb teeth, g is the gap of the teeth.

Then, under the function of the static force, there will be an axial displacement δ. According to the theory of elastic beam, the relationship between the electro static force and the displacement can be expressed as

$$F_{\text{static}} = k\delta \qquad (5.11)$$

where k is the stiffness along the moving direction. k can be calculated by equation (5.9).

If the designed original gap between the contact pair is δ_{design}, the actuator must move at least δ_{design} in order to make the contact pair contact. Combing in equations (5.9), (5.10) and (5.11), the least critical driving voltage can be expressed as

$$v_{\text{critcal}} = \sqrt{\frac{72EIg\delta_{\text{design}}}{n\varepsilon h l^3}} \qquad (5.12)$$

3. The relationship between the driving voltage and the positive pressure

The driving voltage should be increased when the contact of friction pairs are initiated. Electrostatic force should overcome not only the restoring force F_1 caused by bending deformation of six hanging sticks of comb finger actuator B, but also the positive pressure N caused by bending deformation of extended end of actuator A. The relationship is

$$F_{\text{static}} = F_1 + N \qquad (5.13)$$

The value of F_1 can be calculated by equation (5.9). N can be calculated with the relative displacement of cantilever extended end of actuator A in vertical direction. Based on the theory of elastic beam, N can be expressed as

$$N = \frac{3EI_2\delta_2}{l_2^3} = k_{\text{arm}}\delta_2 \qquad (5.14)$$

where k_{arm} is the stiffness of cantilever extended end of actuator A in bending direction; I_2 is the elastic modulus of cantilever extended end of actuator A; l_2 is the length of cantilever extended end of actuator A.

After the contact of friction pair, the relationship between the displacement δ_1 in vertical direction of actuator B's cantilever and the displacement δ_2 in vertical direction of actuator A's cantilever can be expressed as

$$\delta_1 = \delta_2 + 3 \qquad (5.15)$$

Considering these equations, the relationship between the positive pressure N and the driving voltage v is

$$N = k_{\text{arm}} \frac{c_1 v^2 - 3k}{k + k_{\text{arm}}} \qquad (5.16)$$

where c_1 represents parameters related to the size of test facility.

5.3.2.2 Modal analysis and simulation of comb finger actuator

In order to achieve ideal result in the dynamic friction coefficient test and abrasion test of separation decay friction test facility, large relative displacement between friction pairs is necessary. It can be achieved under the resonant state of actuator A. When input frequency of drive power is equal to resonance frequency of actuator, the comb finger actuator is working in the resonant state. Therefore, it's necessary to do the modal analysis of actuator A in order to get the ideal input frequency. The actuator model is shown in Figure 5.1. Modal analysis can be conducted from two aspects of both theoretical calculation and finite element simulation.

The resonance frequency of actuator can be achieved by the following commonly used frequency calculation equation:

$$f = \frac{1}{2\pi}\sqrt{\frac{k}{m_{eq}}} \tag{5.17}$$

where k represents the stiffness of the comb in axis direction, and m_{eq} represents the equivalent mass of actuator.

The stiffness k of actuator in symmetry direction can be calculated by equation (5.9). The calculation of equivalent mass m_{eq} can use the method of conservation of energy. That is, the total kinetic energy is always a constant before and after equivalent. It can be expressed as

$$\frac{1}{2}m_{eq}v_{eq}^2 = \frac{1}{2}m_{rigid}v_{rigid}^2 + \sum_{i=1}^{n}\frac{1}{2}m_i v_i^2 \tag{5.18}$$

The left side of equation (5.18) represents the total kinetic of system after equivalent. The first part of right side of equation (5.18) represents the kinetic of rigid part (including the moving comb, the comb between connections and the beam in the middle) before equivalent. Its second part represents the total energy of flexible parts (six cantilever beams). The symmetry axis of actuator is middle beam and its centroid is on the symmetry axis. Therefore, if set $v_{eq} = v_{rigid} = v$, then equation (5.18) can be expressed as

$$\frac{1}{2}m_{eq}v^2 = \frac{1}{2}m_{rigid}v^2 + \frac{1}{2}\sum_{i=1}^{n}m_i v_i^2 \tag{5.19}$$

As to rigid part of actuator, movement speed of each particle is equal to the centroid, so it's easy to calculate the total energy. As to the flexible structure, the energy can't be calculated directly because the speed of each point is different. If the beam is subdivided into several different parts, the total energy can be calculated using integral method. In order to simplify the calculation, we can choose one pair of cantilevers to calculate because three pairs of cantilevers have the same movement situation. The simplify model can be shown in Figure 5.18. The kinetic energy can be calculated by integral method as follow:

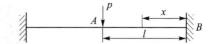

Figure 5.18 Calculation diagram of cantilever structure

$$e_{\text{beam}} = \frac{1}{2}\sum_{i=1}^{n}m_i v_i^2 = \frac{13}{70}m_{\text{beam}}v^2 \tag{5.20}$$

where m_{beam} is the static quality of single cantilever. There are 3 pairs of cantilevers, so the total kinetic is

$$e = 6e_{\text{beam}} = \frac{39}{35}m_{\text{beam}}v^2 \tag{5.21}$$

Comprehensively considering equations (5.21) and (5.19), the equivalent mass of actuator can be expressed as

$$m_{\text{eq}} = m_{\text{rigid}} + \frac{78}{35}m_{\text{beam}} \tag{5.22}$$

According to the given parameters of actuator, equations (5.9) and (5.17) can be synthesized as equation (5.22). The resonance frequency of actuator is about 5 500 Hz.

5.3.2.3 The calculation of friction and positive pressure between friction pair during the testing procedure

Movement of extended cantilever related to static rectangular anchor in horizontal and vertical directions can be test. During the testing procedure, positive pressure and friction between the friction pair can be calculated according to the relative equation and the movement mentioned above. Therefore, it's necessary to derive the relationship among friction, positive pressure and displacement of cantilever in horizontal and vertical directions.

1. The calculation of friction in the testing procedure

During the static and dynamic friction test of separate micro-friction testing device, the value of friction can be calculated by the relative displacement compared with static reference block of extended end of actuator B (point B) in horizontal direction (the calculation model is shown in Figure 5.19). In the process of calculation, actuator B is assumed that the extended end of cantilever (point B) can only produce horizontal displacement. The stiffness of cantilever in horizontal direction is much larger than that of extended end of cantilever. This is the reason of no horizontal direction in root (point A) of extended end of beam in actuator B, so using

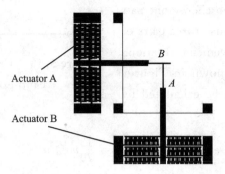

Figure 5.19 Calculating model

the theory of elastic beam, the relationship between friction F_f in test and displacement of point B (δ_h) in horizontal direction can be expressed as

$$F_f = \frac{3EI_B\delta_h}{l_h^3} = k_h\delta_h \qquad (5.23)$$

where k_h represents for stiffness of extended end of actuator B in bending direction. $k_h = \frac{3EI_B}{l_h^3}$, and I_B is moment of inertia of extended end of actuator B.

The value of friction can be achieved easily through the displacement data during testing procedure (image processing method).

2. Calculation of positive pressure during the testing procedure

When DC driving voltage is added on actuator B, friction pairs contact each other. Positive pressure can be calculated with the displacement of extended end B of actuator A in vertical direction. The calculation process is the same with that of friction. Relationship between positive pressure N_{test} and relative displacement δ_v in vertical direction of point B can be expressed as

$$N_{test} = \frac{3EI_A\delta_h}{l_v^3} = k_v\delta_v \qquad (5.24)$$

5.3.3 Technological analysis of structural design

During the testing process of micro friction testing facility, quality of chip production has a great influence on the test result. Therefore, it's necessary to choose an ideal technological process. In this chapter, the technological process of this testing facility is introduced precisely. Optimization procedure and principle of every technological process is also introduced precisely.

Manufacture of micro-friction testing facility on piece is done in the ultra clean room. Main production process uses MEMS technology, stripping process and bonding technology. Before production, mask layouts of two testing facilities using L-edit software are shown in Figure 5.20. Therefore, by using the characteristics of high technology integration and easy-to-batch production, two kinds of testing structures are products on one mask at the same time. In that way, production time is saved and production cost is reduced. From the layout, there are three masks used in production process of testing structure (specific circumstances of each mask are shown in Figure 5.1). Among them, the first piece of mask is used to make the fixed part of the structure (as well as bonded areas); the second piece is used to make actuator and testing part (movable part of testing structure); the third piece is used to make electrode. Specific process is shown in Figure 5.21. First of all, 400 μm thick, $<100>$ orientation 4-inch N-type silicon chip and 4-inch boron silicate heat resistant glass are used as substrate of testing structure. After substrates processing, monocrystalline silicon is corroded by KOH solution. 4 μm's shallow groove is achieved through above operation. This groove is used to form the driver support island and drive side, namely glass bonding area (shown in Figure 5.21 (b)). Then boron is dropped on the silicon to increase the electrical conductivity of the structure, it's shown in Figure 5.21 (c). After the completion of boron's penetration, KOH solution is used to corrode the glass in order to form the shallow groove with the depth of about 120 nm. After shallow groove is formed, sputtering process is used to form Ti/Pt/Au film whose thickness is 100 nm. Then metal electrodes are formed by stripping process shown in Figure 5.21 (f). Next, bonding process is used to put the silicon and glass together. It is shown in Figure 5.21 (g). Actual thickness of silicon wafer is

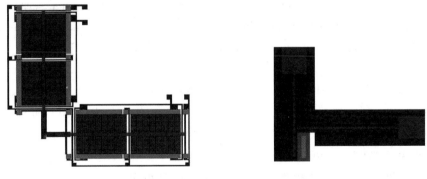

(a) Structure sketch   (b) Enlarged portion of the friction pair

Figure 5.20 Mask sketch of the micro friction system

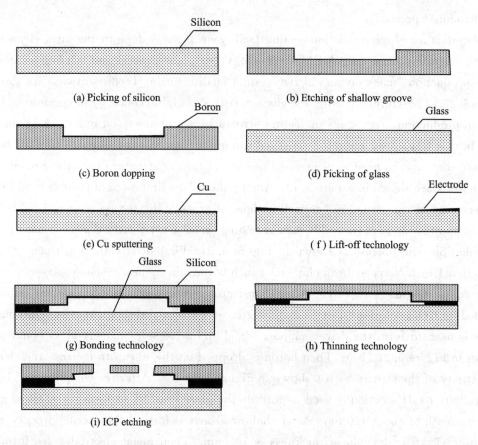

Figure 5.21 Fabrication process of the on-chip micro friction system

greater than its thickness in design. Therefore, KOH solution is used to make the structure thinner. After thinning, the structure is shown in Figure 5.21 (h). Then scribing the glass and sputtering a 500 nm layer on the silicon surface with the material Al film as the mask. After that, ICP etching is used on the silicon to get the final test institution. It is shown in Figure 5.21 (i). Finally, the whole tube core wire is wrapped in a package to complete the whole production process. Figure 5.22 is the pin wiring diagram of test agency under microscope. By the figure, we can see that there's no based electrode fallen off and no warping happened. It shows that the electrode material and substrate glass are combined closely, then the correctness of process is verified. Figure 5.23 shows the partial enlargement (magnification of 1 186 times) under the scanning electron microscopy. From this figure, we can see that the surface of the comb tooth is smooth and steep degree is good. Figure 5.24 shows the figures of two testing institutions under the scanning electron microscopy image—

(a) is the overall design, (b) is an enlargement of the friction pair of SEM photos. It can be seen that the testing institution effect is ideal.

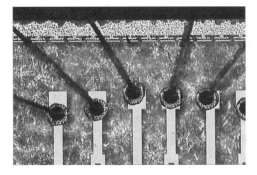

Figure 5.22　Sketch of the sealing

Figure 5.23　SEM sketch of the comb finger

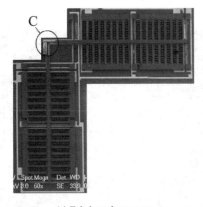
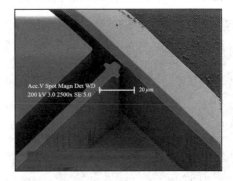

(a) Fabricated structure　　　　　　(b) Enlarged portion of the measuring part

Figure 5.24　SEM sketch of the on-chip friction measuring system

5.3.4 Testing results and data analysis

5.3.4.1 The surface morphology of AFM measurement and data processing

In micro electro mechanical systems, surface topography of friction pair has a great influence on the tribological characteristics. Different surface morphology and measurement conditions (such as test temperature, air humidity etc.) will produce different test results. Therefore, AFM is used to do the morphology test of surface and profile of the friction pair. The tip's scanning area of testing process is 5 μm×5 μm. Along the x and y directions, each tip can acquire 256 points, so the total received data is 65 536. Testing results are shown in Figure 5.25 and Figure 5.26. AFM own processing software can be used to achieve the location and its coordinate value of corresponding height of each testing point. By using MATLAB software, great value of one acquisition point can be processed. From the processing result, it can be seen that the difference between maximum value and minimum value on surface morphology is about 40 nm, while the difference between maximum value and minimum value on profile morphology is about 100 nm. All of these illustrate that there are great differences on profile and surface of testing agency. Therefore it's necessary to do the research on profile friction characteristic of the structure. By using special data processing software of atomic force microscopy and MATLAB software, the morphology parameters of profile of friction pairs are calculated.

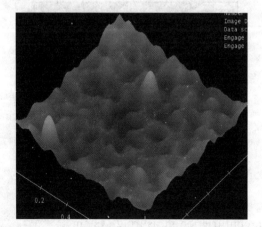
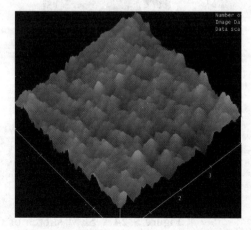

Figure 5.25   AFM sketch of the upper surface   **Figure 5.26**   AFM sketch of the lateral surface

1. The distance calculation of surface peak valley

The distance of surface peak valley represents the attitude difference between the highest and the lowest in measurement range. It can effectively show the maximum relief volume of surface roughness. It can be expressed as

$$R_{max} = R_h - R_l \tag{5.25}$$

where R_h and R_l can be achieved by MATLAB software. Its algorithm is read after the height of each point by using the method of sorting. Though the sorting method, the maximum and the minimum can be get. The R_{max} can be get from the difference of the maximum and the minimum. Application result for R_h is 55.7 m and R_l is −48.8 m. Therefore, the distance between the peak and the valley is 104.5 mm, which can show that the surface of the friction pair using MEMS technology is rough.

2. Average roughness in center direction (S_a) and root mean square roughness (S_q)

These two parameters are most commonly used to describe 3-D surface topography height parameters. Their expressions are shown as follows:

$$S_a = \frac{1}{l_x l_y} \int_0^{l_x} \int_0^{l_y} |z(x,y)| \, dx \, dy \approx \frac{1}{MN} \sum_{i=1}^{M} \sum_{j=1}^{N} |z(x_i, y_j)| \tag{5.26}$$

$$S_q = \sqrt{\frac{1}{l_x l_y} \int_0^{l_x} \int_0^{l_y} z^2(x,y) \, dx \, dy} \approx \sqrt{\frac{1}{MN} \sum_{i=1}^{M} \sum_{j=1}^{N} z^2(x_i, y_j)} \tag{5.27}$$

where l_x and l_y are the scanning lengths of the tip of atomic force microscope in x and y directions. The unit is μm. M and N are the sampling points in x and y directions.

Therefore, under the circumstances of 256 sampling numbers in each direction and 5 μm × 5 μm sampling area, MATLAB software is used to process the morphology of each point. From the above, the average roughness in center direction is $S_a = 10.6$ nm, and root mean square roughness S_q is 13.4 nm.

3. Number of surface peaks and the calculation of its density

The peak value of friction pairs and its density distribution are important parameters to weigh the characters of surface morphology. In this subsection, judgement method of peak-to-peak value is similar with that of eight points in the image processing. The difference of these two is that coordinate test data of the point under this algorithm is the height data of morphology, but not the gray level of pixel. Processing method is still picking one point in the collection point, $z(i,j)$, the surrounding eight points (shown in Table 5.2) and their corresponding height value. If the coordinate height value of the point is greater than that of the other eight points, the point we picked first is the peak of wave. Then, the height value of the peak will be recorded and stored in a matrix using computer software.

Table 5.2 Method to look for the surface peak

$Z(i-1,j+1)$	$Z(i,j+1)$	$Z(i+1,j+1)$
$Z(i-1,j)$	$Z(i,j)$	$Z(i+1,j)$
$Z(i-1,j-1)$	$Z(i,j-1)$	$Z(i+1,j-1)$

The density of surface peak refers to the number of peak on unit sampling area. Generally speaking, the bigger the density is, the larger the contact area will be, and the smaller the stress of contact peak will be. It can be expressed as

$$S_{ds} = \frac{(\text{Number of summits})}{(M-1)(N-1)\Delta x \Delta y} \quad (5.28)$$

Peak number is 1 142 in test area of 5 μm×5 μm by using MATLAB software. From that, density of peak can be calculated as 4.568×10^{13} m^{-2}.

4. Peak-to-peak height average value and standard deviation

By using MATLAB software, the amount of peaks can be calculated. At the same time, the software can also record the height of each peak h_i. So the average height of peak can be expressed as

$$\bar{h} = \frac{\sum_{i=1}^{n} h_i}{n} \quad (5.29)$$

where n stands for the amount of peaks. Its standard deviation can be expressed as

$$\sigma = \sqrt{\frac{\sum_{i=1}^{n}(h_i - \bar{h})^2}{n}} \quad (5.30)$$

From above, the average value of peak is 8.5 nm. Standard deviation is 12 nm, which indicates great variation of wave peak.

5. Arithmetic mean value S_{sc} of curvature of surface peak

Arithmetic mean value S_{sc} of curvature of surface peak on sampling area can be expressed as

$$S_{sc} = -\frac{1}{2} \cdot \frac{1}{n} \sum_{k=1}^{n} \left[\frac{\partial z^2(x,y)}{\partial x^2} + \frac{\partial z^2(x,y)}{\partial y^2} \right] \quad (5.31)$$

where n is the amount of wave peaks. For the measured results, curvature value of wave peak can be calculated by 3 points method as follow:

$$S_{sc} = -\frac{1}{2}\frac{1}{n}\sum_{k=1}^{n}\left(\begin{array}{c} \dfrac{z(x_{p+1}, y_q) + z(x_{p-1}, y_q) - 2z(x_p, y_q)}{\Delta x^2} \\ + \dfrac{z(x_{p+1}, y_q) + z(x_{p-1}, y_q) - 2z(x_p, y_q)}{\Delta x^2} \end{array}\right) \qquad (5.32)$$

By using MATLAB software, arithmetic mean value of the curvature of surface peak is 78 nm.

5.3.4.2 Friction character test of testing institutions

1. Testing principle

On a chip, testing principle of micro friction testing facility is shown in Figure 5.27 and physical figure is shown in Figure 5.28. Friction character test of friction pair is done in ultra-clean chamber (thousand level, room temperature is 23 ℃, air humidity is 21%). The whole experiment system is composed of power supply, friction testing agency, optical microscope (magnification of objective is 100 times), CCD and its connected computer. Power is divided for two purposes. The first one is applying DC voltage to actuator which is placed vertically. This voltage can make the actuator move in vertical direction along and provide positive pressure to the friction test. The other purpose is to provide a series of gradually increasing DC voltage to actuator in horizontal direction by manual operation (process of static friction coefficient test). Or it can provide certain bias of the sine wave voltage (process of dynamic coefficient friction test or abrasion test). The movement condition of friction pair during the testing procedure is magnified by optical microscope and

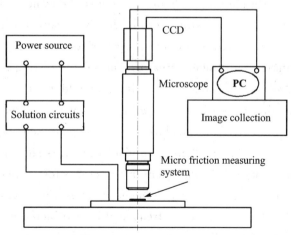

Figure 5.27 The testing system

it's acquired by CCD. The acquired data is transferred to the computer through data acquisition card and the data is processed by computer software. Then, MATLAB software is used to process the image acquired and produce friction data required. Before the test, CCD pixels of microscope is calibrated by raster with given parameters, shown in Figure 5.29. From the above, the displacement resolution of objective of microscope with the magnification of 100 is 10.2 pixels/μm. Therefore, the displacement resolution during the testing procedure is 0.1 μm.

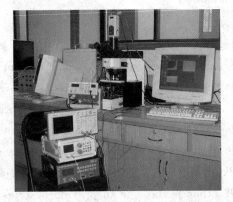

Figure 5.28 A real testing system

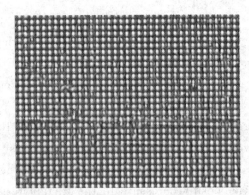

Figure 5.29 Sketch of the optical grating under 100× microscope

2. Resonance frequency test of comb finger actuator

During the testing procedure of dynamic coefficient and wear condition of friction pairs, the frequency of input offset sine wave should be almost the same with resonance frequency of actuator in order to get ideal testing result. In that way, the actuator will work in resonance state. Therefore, before friction characteristic test, resonance frequency test should be done on a single comb finger actuator. During the testing procedure, bias voltage offset and peak-to-peak value of sine wave voltage should be 17 V. That is $v_A = 17 + 17\sin \omega t$. Then, when the value of driving voltage is fixed, voltage frequency applied on actuator is increasing gradually. By calculating every displacement of the input frequency, the relation curve between input frequency and displacement is shown in Figure 5.30. From the figure, it can be seen that the displacement of actuator is increasing with the increasing of voltage frequency of actuator. The displacement reaches the maximum when the input frequency is about 5 700 Hz. If the input frequency is also increasing, the displacement of actuator will suddenly decrease. This indicates that 5 700 Hz is the resonance frequency of actuator. It's almost the same with the theoretical result 5 590 Hz.

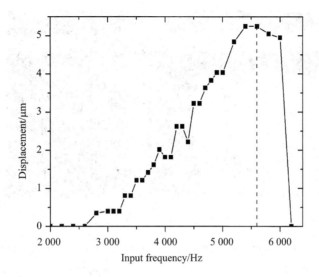

Figure 5.30 Relationship between the input frequency and the displacement

3. Test of static friction coefficient

There are two actuators in vertical and horizontal directions, respectively. The actuator in vertical direction is applied with constant DC voltage. Under this circumstance, static friction coefficient of friction pairs can be achieved with a series of gradually increasing DC voltage on horizontal actuator. Status of friction pair under optical microscope during testing process can be shown in Figure 5.31. During the test, 40 V DC voltage is applied on actuator B in vertical direction (critical contact value of actuator is 30 V according to the test). Corresponding contact pressure can be measured with testing of vertical displacement of beam's extended end of actuator A with equation (5.24). After the contact of friction pairs, a series of gradually increasing DC voltage (0~60 V) are added on actuator A. By observing relative displacement between each friction pair under the corresponding voltage, the static friction properties can be studied.

With the increasing of driving voltage, the displacement of contact part of vertical beam in horizontal direction and the deformation energy increase. It's shown in Figure 5.32. The friction corresponded with deformation of beam can be calculated by using the relative equation and the horizontal displacement of the end of vertical beam.

Therefore, with the increasing of deformation energy, the friction between the friction pair increases. Then, the deformation of beam continues rising because the increasing of driving voltage. When restoring force caused by deformation is bigger than friction between

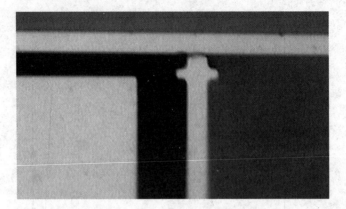

Figure 5.31 Measuring principle of the static friction coefficient

the friction pair again, the beam will rebound. As a result, the static friction at the moment that top of beam rebounds is the maximum static friction which friction pairs can provide. The ratio (0.9) between the corresponding friction and positive pressure is static friction coefficient between the friction pair.

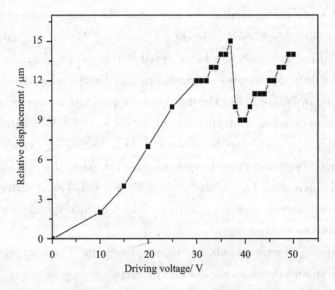

Figure 5.32 Measuring curve of the static friction coefficient

4. Test of dynamic friction coefficient

A sine wave with bias DC voltage is applied on actuator A after positive pressure added on actuator B. Dynamic friction coefficient of the friction pair can be tested by the above procedure. During the test, resonance frequency of actuator is equal to the input frequency of sine wave. The resonance frequency is about 5 500 Hz. The amplitudes of sine wave and DC offset voltage are all 20 V. The figure of friction pairs contacting each other in resonance situation is Figure 5.33. At the same time, dynamic friction coefficient can be achieved by testing the relative displacement of the top of vertical beam in horizontal direction. And the corresponding positive pressure can be tested through the displacement of the top of vertical beam and horizontal beam in vertical direction. Table 5.3 shows test results of above parameters under driving voltage of 35 V, 40 V, 50 V and 60 V on actuator in vertical direction.

Figure 5.33 Measuring sketch of the dynamic friction coefficient

Table 5.3 Pertinent measuring data of the micro friction

Driving voltage/V	Normal pressure/μN	Friction force/μN	Dynamic friction coefficient
0	0	0	—
35	76.83	18.1	0.24
40	128.05	42.2	0.33
50	204.88	72.4	0.35
60	307.32	78.4	0.26

According to the traditional theory of tribology, the friction value is proportional to the positive pressure added on friction pairs. The relationship can be expressed as

$$f = \mu N \tag{5.33}$$

Parameter μ in above equation is a constant, and it's called friction coefficient. From Table 5.3 and Figure 5.19, it can be seen that dynamic friction coefficient of friction pair is no longer a constant. It's changing with the change of the positive pressure applied on the friction pair. Visibly, the test result is consistent with the traditional law of friction.

5. Explanation of test results

According to the Hertz contact theory, when two cylinders contact with the positive pressure W applied, half of bandwidth b in contact area can be represented as equation (5.34). Model of Hertz contact theory is shown in Figure 5.34.

$$b = \left(\frac{8}{\pi} \frac{WR}{lE'}\right)^{\frac{1}{2}} \quad (5.34)$$

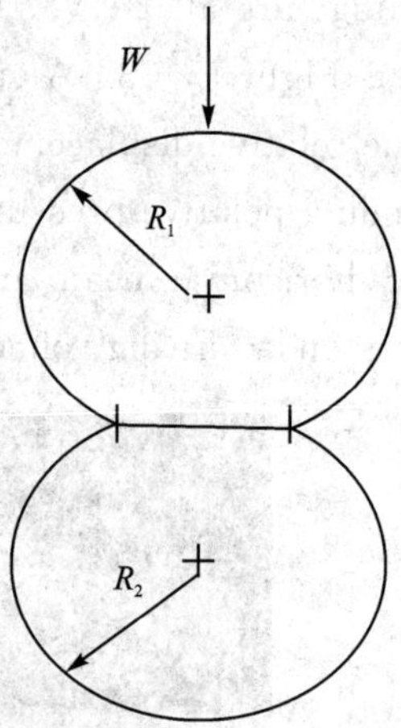

where, l represents length of contact area, R and E' represent equivalent radius and equivalent elastic modulus respectively. They can be expressed as $\frac{1}{R} = \frac{1}{R_1} + \frac{1}{R_2}$ and $\frac{1}{E'} = \frac{1-v_1^2}{E_1} + \frac{1-v_2^2}{E_2}$. E_1 and E_2 represent

Figure 5.34 Model of the Hertz contact theory

elastic modulus of two kinds of materials respectively. v_1 and v_2 represent the Poisson's ratios of two kinds of materials respectively. When friction pairs have the same material and two contact friction pairs have the shape of cylinder and flat, the above two equations can be expressed as $\frac{1}{R} = \frac{2}{R'}$ and $\frac{1}{E'} = 2 \cdot \frac{1-v^2}{E}$.

Therefore, Hertz contact stress distribution equation on contact surface is

$$p = p_0 \left(1 - \frac{x^2}{b^2}\right)^{\frac{1}{2}} \quad (-b < x < b) \quad (5.35)$$

The maximum Hertz stress is

$$p_0 = \frac{4}{x} p_c = \frac{2}{\pi} \frac{W}{bl} \quad (5.36)$$

Average contact stress is

$$p_e = \frac{W}{2bl} \quad (5.37)$$

Soviet scholar кратедьскии[67] and some other people hold the view that sliding friction is the procedure of overcoming the mechanical gearing and molecular attraction of rough surface peak. It can be expressed as

$$F = \tau_0 S_0 + \tau_m S_m \quad (5.38)$$

where S_0 and S_m represent the area of molecular and mechanical action respectively, τ_0 and τ_m

are the friction in unit area produced by molecular and mechanical action. The above equation can be also expressed as

$$F = \alpha A + \beta W \tag{5.39}$$

which is called binomial law of friction. A in equation (5.39) is physical function load and W is normal load. The friction coefficient can be expressed as

$$f = \frac{\alpha A}{W} + \beta \tag{5.40}$$

where f is changing with the change of A/W. Under macroscopic condition, molecular interatomic force is much smaller than the force caused by mechanical surface topography. So the first item in equation (5.40) can be ignored. The only thing should be considered is coefficient friction caused by surface mechanical function. However, it's a constant. Therefore, under macroscopic condition, the law of tribology conforms to the law of Amontons. That is, the friction is proportional to the positive pressure. When the size of the contact pair is at micron level or even smaller, the intermolecular force becomes larger when the size of contact pair becomes smaller. At that time, the intermolecular force can't be ignored any more, so friction coefficient is not a constant any more. It's changing with the change of positive pressure.

5.3.5 Research and test of wear problem of MEMS devices

Under micro-scale condition, the wear properties of MEMS device have great difference with macroscopic condition. For specific, the wear material and macro conditions are not always the same with the increase of surface energy. Therefore, after the coefficient test of friction pair under static and dynamic conditions, the wear condition is tested under microscope.

In order to test the surface composition change of friction pair before and after wear, the composition of friction pair is detected by scanning electron microscopy before test. The energy spectrum diagram is shown in Figure 5.35. From the figure, it can be seen that the major component of

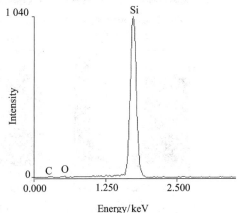

Figure 5.35 Energy spectrum diagram of the constituent for the friction pair

friction pair is silicon, and carbon and oxygen are less in friction pair while most of those is the legacy component in manufacturing operation. Related parameters of driving power and some data in wear test procedure are shown in Table 5.4. From the table, relative displacement between friction pairs can reach 5 742 m.

Table 5.4 Parameters index during the course of the wear and tear

Measuring parameters	Frequency of the driving voltage/Hz	Working time/h	Wording period	Relative displacement between the friction pair/μm	Moving distance/m
Value	2 900	55	574 200 000	10	5 742

In order to monitor the wear process more accurate, optical microscope is used to test every stage of the shape of friction pairs, which can be shown in Figure 5.36. From the figure, it can be seen that the wear condition of friction pair is serious when the wear time reaches 33 hours. However, at the wear time of 55 hours, both the side and the surface of friction pair are seriously wore. During the testing procedure, the friction coefficient of every stage is tracked and it is increasing gradually because of wear. Therefore, in order to keep the friction pairs staying in touch, the positive pressure is adjusted in different testing stages

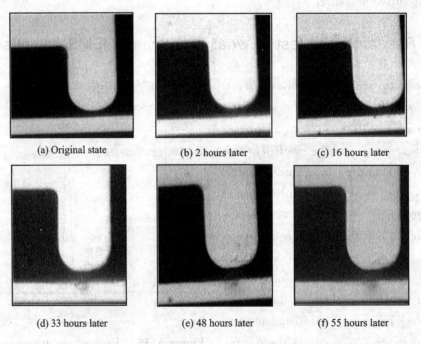

(a) Original state (b) 2 hours later (c) 16 hours later

(d) 33 hours later (e) 48 hours later (f) 55 hours later

Figure 5.36 Shape changing of the contact pair

(this is realized by increasing the driving voltage of actuator B).

Driving voltage and friction coefficient in every stage is shown in Table 5.5. From the table, it can be seen that friction coefficient is the biggest at the beginning. It shows that friction coefficient is still in running-in status. With the wearing procedure, friction coefficient decreases and reaches 0.09 finally. This shows that friction pairs are in stable wear stage. After wearing test, scanning electron microscopy images (magnification is 12 000 times) of the side and the surface of friction pairs are shown in Figure 5.37. From the figure, it can be seen that the friction pair is wore seriously and some grains appeared on the surface of friction pair.

Table 5.5 Change of the dynamic friction coefficient

Normal pressure/V	Dynamic displacement with no loading/μm		Dynamic displacement with loading applied/μm	Relative displacement/μm	Friction force/μN	Dynamic friction coefficient
At beginning (40 V applied)	0	4.7	4.21	0.49	27.44	0.41
	16 hours later	4.7	4.43	0.27	15.12	0.23
16 hours later (45 V applied)	At beginning (40 V)	4.7	4.27	0.43	24.08	0.25
	9 hours later	4.7	4.37	0.33	18.48	0.19
	17 hours later	4.7	4.43	0.27	15.2	0.16
33 hours later (48 V applied)	At beginning (40 V)	4.7	4.27	0.43	24.1	0.18
	8 hours later	4.7	4.3	0.4	22.4	0.17
	17 hours later	4.7	4.5	0.2	11.2	0.09
	22 hours later	4.7	4.5	0.2	11.2	0.09

At the same time, under SEM, the comparison between shapes of abrasive particles on the surface and the side of friction pairs is made. The result is shown in Figure 5.38. From this figure, it can be seen that abrasive particles gather into piles on surface and side of friction pairs. But not like particles evenly dispersed on the surface of friction pairs in macro condition. According to the interpretation of reference [59], the formation of this phenomenon may be due to the effect of water vapor in the air. The vapor may increase the surface force of abrasive particle, so that abrasive particle will gather together. From shape test of abrasive particle, density of abrasive particle on side of friction pair is big. However, density on surface of abrasive is small. Abrasive particles on side are generated into grain due to the procedure of repeated compression on abrasive caused by friction pair. At the same time, test on component of abrasive particle of friction pair is made under electron microscopy after experiment completes. The spectrum of composition is shown in Figure

5.39. From the spectrum, it can be seen that the oxygen content on friction pair is increased greatly after wear test. This shows that during the wear test, silica is generated because of oxidation reaction. This is maybe due to the material's oxidation caused by high temperature of friction pairs during the procedure of friction. Visibly, wear process is a complex process. During the process, not only physical change but also chemical reaction occurs.

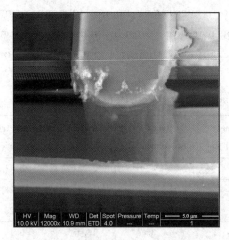
(a) SEM sketch on the lateral surface

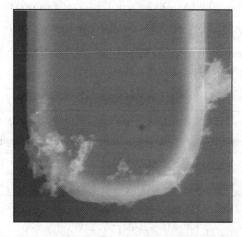
(b) SEM sketch on the top surface

Figure 5.37 SEM sketches of the friction pair after the testing of wear and tear

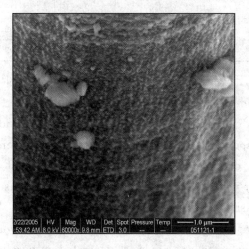
(a) Shape of friction particles on the lateral surface

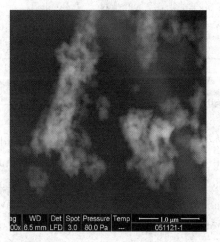
(b) Shape of friction particles on the top surface

Figure 5.38 Comparison of friction particles on the top and lateral surfaces

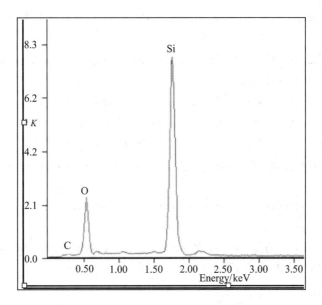

Figure 5.39 SEM sketch of the constitute for the friction particles

5.4 Summary

In this chapter, the research status and the development trend of MEMS tribological properties at home and abroad are introduced in detail. Tribological properties, test methods and results of MEMS devices are introduced using the engineering examples.

References

[1] Huang Xinbo, Wang Weidong. The State of the Art in Technologies and Applications of MEMS[J]. Mechanical Science and Technology,2003. (22):21-24.

[2] David J Nagel. Design of MEMS and Microsystems. Proceedings of the SPIE, 1999, 3680(1):20-29.

[3] Husak, Miroslav. System of Models for MEMS Design and Relization[J]. WSRAS Transaction on Systems, 2005,4(3):175-184.

[4] Varadan V K, Vinoy K J. Application of MEMS in Microwave and Millimeter Wave Systems. Proceedings of SPIE,2001,4236:179-187.

[5] Bouchaud, Jeremie, Kharas, et al. RF MEMS: Status of the Industry in 2004,

Application Roadmap and Market Forecasts[J]. Sensors and Actuators, A: Physical, 2003,104(1):1-5.

[6] Fujita, Hiroyuki. MEMS/MOMES Application to Optical Communication. Proceedings of SPIE, 2001,4559:xxi-xxvii.

[7] Matthew A Beasley, Samara L Firebaugh. MEMS Thermal Switch for Spacecraft Thermal Control. Proceedings of SPIE,2004,5334:98-105.

[8] Rossi C, Do conto T, Esteve D. Design, Fabrication and Modelling of MEMS-based Microthrusters for Space Application[J]. Smart Materials and Structures, 2001,10(6): 1156-1162.

[9] Ong, Zhiyang, Al-Sarawi. Surgical Application of MEMS Devices. Proceedings of SPIE, 2005,5649:849-860.

[10] Liu Ran, Zhou Zhaoying, Wang Xiaohao. The Application of MEMS Microneedles in Biomedicine. Porceedings of the International Symposium on Test and Measurement, 2003,1:57-60.

[11] Kakushima Kuniyuki, Fujita Hiroyuki. MEMS Application to Characterization of Field Emitters and Bio Molecules. Proceedings of SPIE, 2004,5455:82-88.

[12] Goel Malti. Recent Development in Electroceramics: MEMS Applications for Energy and Environment[J]. Ceramics International, 2004,30(7):1147-1154.

[13] Bushan B. Tribology on the Macroscale to Nanoscale of Microelectromechanical Systems Materials: a Review. Proceedings of the Institution of Mechanical Engineers, Part J. Journal of Engineering Tribology, 2001,215(1):1-18.

[14] Willams J A. Friction and Wear of Rotating Pivots in MEMS and Other Small Scale Devices[J]. Wear,2001,251:965-972.

[15] Wang Qingliang, Ge Shirong. Progress of Research on Nanotribology of Microelectromechanical Systems[J]. Lubrication Engineering,2003(3):88-91.

[16] Trimmer W S N, Gabriel K J. Design Considerations for A Practical Electrostatic Micro-motor[J]. Sensors and Actuators,1987,11(2):189-206.

[17] Gatzen H H, Beck Michael. Tribological Investigations on Micromachined Silicon Sliders[J]. Tribology International 2003,36:279-283.

[18] Bharat Bhushan, Senlor Member. Nanotribology and Nanomechanics of MEMS Devices. Proceedings of the IEEE Micro Electro Mechanical Systems,1996:91-98.

[19] Sriram Sundararajan, Bharat Bhushan. Topography-induced Contributions to Friction Force Measured Using an Atomic Force/Friction Microscope[J]. Journal of Applied Physics, 2000,88(8):4825-4831.

[20] Schmidt Martin, Wortmann Andreas, Luthje Holger, et al. Novel Equipment for Friction Force Measurement on MEMS and Micro Components. Proceedings of SPIE, 2001,4407:158-163.

[21] Gatzen H H, Beck M. Wear of Single Crystal Silicon as a Function of Surface Roughness[J]. Wear,2003(254):907-910.

[22] Bandorf R, Lüthje H, Wortmann A, et al. Influence of Substrate Material and Topography on the Tribological Behaviour of Submicron Coatings[J]. Surface and Coatings Technology,2003(174):461-464.

[23] Chen Quanfang, Carman Greg P. Microscale Tribology(friction) Measurement and Influence of Crystal Orientation and Fabrication Process. Proceedings of IEEE: 657-661.

[24] Tas Niels, Sonnenberg Tonny, Jansen Henri, et al. Stiction in Surface Micromachining[J]. Journal of Micromechanics and Microengineering, 1996,6(4):385-397.

[25] Tas N R, Gui C, Melwenspoek. Static Friction in Elastic Adhesive MEMS Contacts, Models and Experiment. Proceedings of the IEEE Micro Electro Mechanical Systems (MEMS), 2000:193-198.

[26] Tas N R, Gui C, Melwenspoek. Static Friction in Elastic Adhesion Contacts in MEMS. Journal of Adhesion Science and Technology, 2003,17(4):547-561.

[27] Deladi S, Boer M J de, Krijnen G, et al. Innovative Process Development for a New Micro-tribosensor Using Surface Micromachining[J]. Journal of Micromechanics and Microengineering,2003,13:17-22.

[28] Senft Donna Cowell, Dugger Michael T. Friction and Wear in Surface Micromachined Tribological Test Devices. Proceedings of the SPIE,1997,3224:31-38.

[29] Tanner Danelle M, Dugger Michael T. Wear Mechanisms in A Reliability Methodology. Proceedings of SPIE,2003,4980:22-40.

[30] Dugger Michael T, Hohlfelder Robert J, Peebles Diane E. Degradation of Monolayer Lubricants for MEMS. Proceedings of the SPIE,2003,4980:138-150.

[31] Alsem D H, Stach E A, Muhlstein C L, et al. Utilizing On-chip Testing and Electron Microscopy to Study Fatigue and Wear in Polysilicon Structural Films[J]. Materials Research Society,2004,1821:1-6.

[32] Guo Zhanshe, Meng Yonggang, Hao Wu, et al. Measurement of Static and Dynamic Friction Coefficient of Sidewalls of Bulk-microfabricated MEMS Devices with An On-chip Micro-tribotester[J]. Sensors and Actuators A-physical, 2007,135:863-869.

[33] Lim M G, Chang J C, Schultz D P, et al. Polysilicon Microstructures to Characterize

Static Friction. Proceedings of the IEEE Micro Electro Mechanical Systems-An Investigation of Micro Structures, Sensors[J], Actuators, Machines,1990,82-88.

[34] Komvopoulos K. Surface Engineering and Microtribology for Microelectromechanical Systems[J]. Wear,1996,200:305-327.

[35] Ashursta W Robert, Yaua Christina, Carraro Carlo. Alkene Based Monolayer Films as Anti-stiction Coating for Polysilicon MEMS[J]. Sensors and Actuators A,2001,91:239-248.

[36] Lumbantobing A, Komvopoulos K. Static Friction in Polysilicon Surface Micromachines[J]. Journal of Micrelectromechanical Systems,2005,14(4):651-653.

[37] Ernest J Garcia, Jeffry J Sniegowski. Surface Micromachined Microengine[J]. Sensors and Actuators A,1995,48:203-214.

[38] Miller S L, Sniegowski J J, Vigen G La. Friction in Surface Micromachined Microengines. Proceedings of SPIE,1996,2722:197-204.

[39] Eaton William P, Smith Norman F, Irwin Lloyd, et al. Characterization Technique for Surface-Micromachined Devices. Proceedings of SPIE,1998,3514:431-437.

[40] Tanner Danelle M, Walraven Jeremy A, Irwin Lloyd W et al. The Effect of Humidity on the Reliability of a Surface Micromachined Microengine. Proceedings of the 1999 37 th Annual IEEE International Reliability Physics Symposium:189-197.

[41] Ashurst W Robert, Yau Christina, Carraro Carlo et al. Dichlorodimethylsilane as An Anti-Stiction Monolayer for MEMS: A Comparison to The Octadecyltrichlosilane Self-Assembled Monolayer[J]. Journal of Micromechanical System. 2001,10(1):41-49.

[42] Hankins Matthew G, Resnick Paul J, Clews Peggy J et al. Vapor Deposition of Amino-Functionalized Self-assembled Monolayers on MEMS. Proceedings of SPIE, 2003,4980:238-247.

[43] Patton S T, Cowan W D, Zabinski J S. Performance and Reliability of a New MEMS Electrostatic Lateral Output Motor. Annual Proceedings of Reliability Physics (Symposium),1999:179-188.

[44] Patton Steven T, Cowan Willam D, Eapen Kalathil C et al. Effect of Surface Chemistry on The Triblogy Performance of A MEMS Electrostatic Lateral Output Motor[J]. Tribology Letters,2000,9:3-4.

[45] Smallwood Steven A, Eapen Kalathil C, Patton Steven T et al. Performance Results of MEMS Coated with A Conformal DLC[J]. Wear,2006, 260(11-12):1179-1189.

[46] Chen Chihchung, Lee Chenkuo. Design and Modeling for Comb Drive Actuator with Enlarged Static Displacement[J]. Sensors and Actuators, A: physical, 2004, 115 (2-3): 530-539.

Chapter 6
MEMS testing technology and engineering application

This chapter is mainly based on acceleration, angular velocity, pressure, micro torque testing, micromechanical testing and micro-morphology testing. Some relevant test methods, basic theories and devices for MEMS testing are selected as the study object. The detailed introductions of testing theories and methods on various parameters, and the domestic and foreign development are given. The author also combine with his scientific achievements to help the reader get a further understanding on the theory.

6.1 Acceleration measurement and corresponding sensors

Accelerometer is a force-sensitive sensor for testing the moving carrier acceleration. In 1942, accelerometer was firstly used in inertial navigation and inertial guidance systems. In that year, a German scientist Pella Richmond successfully launched the world's first V-2 Rocket, and in the longitudinal axis of the rocket, installed one integral line accelerometer to control the rocket's engine flameout[1]. As one of the most important inertia instruments, accelerometer has been one of the world's key projects. The early typically accelerometers conclude liquid floated pendulous accelerometer[2], vibrating spring accelerometer[3] and flexure accelerometer[4].

With the development of micro-nanometer technology, a variety of new micro-accelerometers based on integrated circuit technique and MEMS fabrication techniques are emerging, and these micro-accelerometers are widely used in aeronautics, astronautics and national defense, and quickly spread to civilian areas[5], such as automotive electronics[6], mobile phone[7], games[8] and medicine[9] etc.

6.1.1 Working principle of the acceleration sensor and the classification

The schematic diagram of silicon micromechanical accelerometer working principle is shown in Figure 6.1. The transducer's structural support is composed of spring, mass and damping structure. When in the acceleration effect y-direction applied on transducers, according to Newton's law, the inertial force F produced by the mass can be expressed as

$$F = ma \tag{6.1}$$

where m is the quality of the mass, and a is the acceleration.

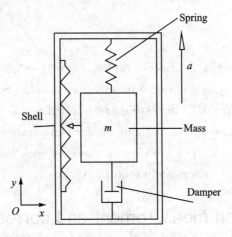

Figure 6.1 Working principle schematic diagram of accelerometer sensor

According to equation (6.1), the relationship between the acceleration and the inertial force is

$$a = \frac{F}{m} \tag{6.2}$$

To the fabricated acceleration sensor, the value of mass m is fixed. If F is obtained, the acceleration can be obtained by equation (6.2). Meanwhile the inertial force will cause effects on the device. One effect will cause the displacement in the structure. Therefore, by testing the displacement caused by the inertial force, and calculating under the corresponding theory, the value of the acceleration can be obtained then. This sensor is called as analog accelerometer. According to Hooke's law, when the force is applied to the elastic structure, the equation can be expressed as

$$F = k_{\text{spr}} x \tag{6.3}$$

Combine equations (6.2) and (6.3), and get the equation expressing the acceleration and displacement as follow:

$$a = \frac{k_{spr} x}{m} \tag{6.4}$$

For such kind of sensors, so long as the displacement is measured, the acceleration will be obtained. We classify this type of sensor according to their sensitive mechanism into capacitive accelerometer, piezoresistive, piezoelectric, electromagnetic and the tunneling current accelerometer et al.

The other effect of the device is the change of resonant frequency for the sensitive element. Thus the other method of mechanical testing on sensors is accomplished through its own mechanical properties, such as the change of sensitive element's resonant frequency caused by the change of its own mechanical properties. The concrete way to realize it is to add sensitive element (shown in Figure 6.2) on both ends of the mass.

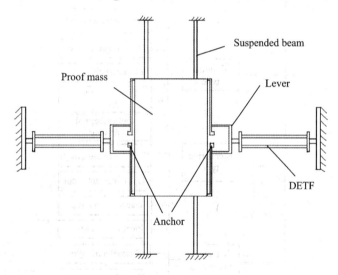

Figure 6.2 Resonant silicon micromechanical accelerometer working principle

when the inertial force (transverse direction) is applied on the tuning fork, one effect is the resonance frequency will be changed (details are introduced in Chapter 2). Their relationship can be expressed as

$$f = f(a) \tag{6.5}$$

where f is the resonant frequency of sensitive element. We can get the acceleration by testing the change of resonance frequency. We call this kind of sensor as resonant silicon micromechanical accelerometer. Because this kind of sensor measures the acceleration by measuring the change of the resonant frequency, it has the advantages of high testing precision, good stability, and interfacing with computers and digital output easily. The sensor is the important research direction in high-performance accelerometer field.

According to the different testing methods, the silicon micromechanical accelerometer can be classified into capacitance type, piezo resistive type, piezoelectric type and resonant type and so on, as shown in Figure 6.3.

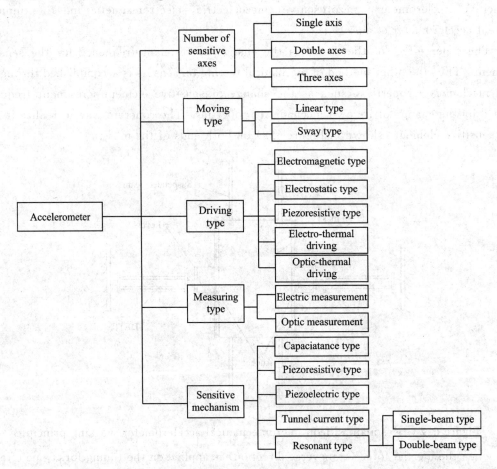

Figure 6.3 Silicon micromechanical accelerometer classification

6.1.2 Capacitive silicon micromechanical accelerometer

Capacitive silicon micromechanical accelerometer is designed in comb capacitor structure. When the comb teeth shift caused by external load, the capacitance will change,

Chapter 6 MEMS testing technology and engineering application

and we obtain the acceleration value by testing the capacitance change. Referring to the theories mentioned in Chapter 2, we know that the capacitance is arranged in two ways—area-changing and gap-variable. So sensors can be classified into two types—area-changing acceleration sensor and gap-variable sensor. In actual projects, the sensor is designed in differential structure to reduce the errors.

Here are some representative products, which are produced by ASC from Germany and Kistler from Switzerland, as shown in Figure 6.4. The capacitive accelerometer possesses the advantages that get wide frequency response and large measuring range, and is one kind of silicon micromechanical accelerometer widely used in current years.

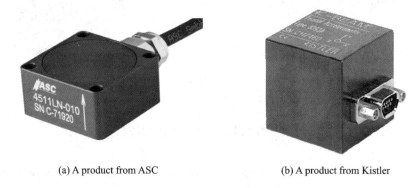

(a) A product from ASC　　　　　　　　(b) A product from Kistler

Figure 6.4　Typical capacitive accelerometers

6.1.3　Piezoresistive silicon micromechanical accelerometer

The piezoresistive silicon micromechanical accelerometer is based on semiconductor materials piezoresistive effect. The displacement changes generate the resistance of the sensitive element change, so we can detect the resistance change to obtain the acceleration (details are introduced in Chapter 3). The piezoresistive property is easily affected by temperature change, so generally, the bridge circuit is used in actual projects, and will convert the resistance change to the change of voltage or current.

The greatest advantage of piezoresistive accelerometer is easy to integrate and produce. And because this type of sensors obtain the acceleration by the semiconductor piezoresistive effect, the properties are vulnerable to temperature change. The low measurement accuracy is the weakness, so piezoresistive accelerometer is used in the low measurement accuracy field. The main application fields of piezoresistive accelerometer are the ignition device in military, the dynamic measurement device in flight, the displacement detection in fail-safe system, and the study of mechanics in auto industry. The representative products are from American companies Endevco and MEAS, as shown in Figure 6.5.

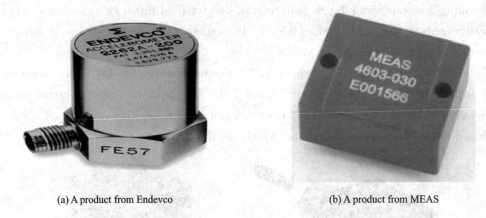

(a) A product from Endevco   (b) A product from MEAS

Figure 6.5 Piezoresistive accelerometers

6.1.4 Piezoelectric micromechanical accelerometer

Piezoelectric micromechanical accelerometer is based on the piezoelectric material piezoelectric effect to accomplish the testing of the acceleration (details are introduced in Chapter 3). Its structural schematic diagram is shown in Figure 6.6. The piezoelectric accelerometer is comprised of mass block, hard spring, piezoelectric chip and base. The object moves, then the accelerometer moves, and it will generate very small inertia because the mass is too light compared to the high stiffness of the hard spring. So the acceleration of the mass is almost equivalent to the acceleration of the object, and the mass block sets up the inertia force which is proportional to the acceleration. The inertial force acts on the piezoelectric substrate, and generates the charge or voltage which are proportional to the acceleration. Then we can obtain the acceleration by testing the change of charge or voltage. Here are the typical piezoelectric accelerometer products from Company Endevco and MEAS from the USA, as shown in Figure 6.7.

The most outstanding feature of the piezoelectric accelerometer are the high frequency response, in micro volume and mass, widely measuring range, and operated in a large temperature range. But because of the influence of the leakage current and thermoelectric effect, this kind of accelerometers get the weakness of large temperature drift and poor low frequency response, so the accelerometers are not for the steady state measurement. These

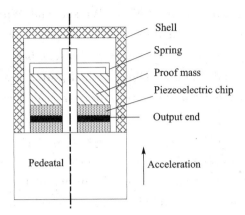

Figure 6.6   Schematic diagram of piezoelectric accelerometer

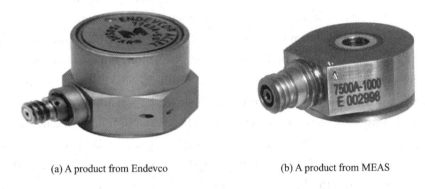

Figure 6.7   Piezoelectric accelerometers

defects limit the application range of this kind of accelerometer.

6.1.5   Resonant silicon MEMS accelerometer

As introduced in Subsection 6.1.1, the resonant silicon MEMS accelerometer is one kind of force sensor. It utilizes the stress-frequency effect to accomplish the acceleration testing. The stress-frequency effect means that the external stress that acts on the resonant sensor will change the internal stress state, and the resonant frequency of the sensitive element will also change. The external stress can be external force, inertial force, vibration and pressure etc.

In 1964, Willis and Jameson put the stress frequency effect into use firstly. They created the first principle prototype of the accelerometer with the direct frequency output using the quartz material. And in the 1970s, Watson and William also had done some research on the accelerometer with the direct output of frequency.

After coming into 1980's, all kinds of micro sensors based on the integrated circuit technology and the MEMS fabrication technique, had gotten a great development. At the same time, the direct frequency output accelerometer had been a new focus in science research.

In 1979, the USA Stanford University took advantage of the MEMS fabrication technique and the silicon, developed the open loop direct frequency output accelerometer. In 1989, Satchell D. W. and Greenwood J. C. by the means of thermal excitation and voltage detecting, designed the silicon direct frequency output accelerometer with three beams. In 1990, Chang C. S. designed the direct frequency output accelerometer that can be sensitive in two directions on using of two-beam structure, then applied the patent. In 1991, the Draper Laboratory used the tuning-fork structure in the resonant sensor, and developed the quartz direct frequency output accelerometer. In 1991, 1994 and 1996, Cheshmehdoost A. did the theoretical analysis and experimental research on the tuning-fork of the direct frequency output accelerometer. In 1997, University of California-Berkeley developed the direct frequency output accelerometer with the tuning-fork as the sensitive structure.

When came into this century, research on the direct output frequency accelerometer have developed rapidly, and these research institutes, such as Draper Laboratory, MIT, University of California-Berkeley, Honeywell and Freestyle, are at the peak of the research. Europe and Japan also get the high research level.

The typical product is RBA500 developed by Honeywell, as shown in Figure 6.8. Here are some parameters of these serial products: the range of accelerometer is ± 70 g, the scale factor is 80 Hz/g, the resolution ratio is <1 μg.

Figure 6.8 The silicon micro-machined resonant accelerometer RBA500 produced by Honeywell

6.2 Angular speed measurement and corresponding sensors

The angular speed measurement of the moving objects is mainly achieved by gyroscope. This sensor and the acceleration sensor assemble the IMU (Inertial Measurement Unit), which can detect the moving attitude of the moving objects. With the development of MEMS technology, the MIMU (Micro Inertial Measurement Unit) system with the advantages of miniaturization, low power consumption, high integration and low cost has become the main development direction of inertial systems. Being the most important key element of the inertial system, miniaturization of the gyroscope has become its main research direction. They have been widely used in aviation, spaceflight, navigation, military and other fields from its coming out. And the development of gyroscope is always a national key project[17-20].

6.2.1 Working principle

Simple working principle of the silicon MEMS gyroscope is shown in Figure 6.9. The gyroscope is mainly made up with two sets of flexible supporting structures, two sets of damping structures and a mass proof. These two flexible support-damping structures are arranged in perpendicular direction. The system can be considered as a 2-D vibration system. Generally, the mass will make the reciprocating motion in y direction under the function of driving force (frequency equals to the resonant frequency in y direction). When an angular velocity along the z-axis is imposed on the system, because of the Coriolis Effect, there will be a Coriolis force in x-axis direction. It can be expressed as

$$F_C = 2\, mv \times \Omega \tag{6.6}$$

where m is the mass, v is the y direction linear velocity, and Ω is angular velocity.

When the linear velocity perpendicular to the angular velocity and the velocity and it is sine type, equation (6.6) can be expressed as

$$F_C = 2\, mV\Omega \sin \omega_n t \tag{6.7}$$

where V is the moving peak value, and ω_n is the angular velocity.

It is concluded from the equation that the Coriolis force is proportion to the mass. So, the detection of the angular velocity is equal to the detection of Coriolis force. According to the sensing mechanism, the micromechanical gyroscope can be classified into Capacitance type, piezoresistive type, piezoelectric and resonant etc.

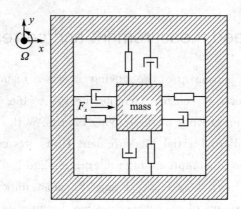

Figure 6.9 Working principle of the silicon MEMS gyroscope

6.2.2 Development of MEMS gyroscope

Research of the silicon MEMS gyroscope was initiated from late eighties of the 20th century. The MEMS technology based on the IC process began to be applied to the manufacture of the sensor, which greatly promoted the development of MEMS gyroscopes and the silicon MEMS gyroscope began to develop. MEMS gyroscope works in the same principle with the vibrating piezoelectric crystal gyroscope. However, the processing methods, device characteristics and application prospects have a significant difference to the piezoelectric crystal gyroscope. The MEMS gyroscope process uses the semiconductor processing technology fabricated on monocrystalline silicon or quartz, and the associated electronic circuits will be integrated on the same chip. This combination of microelectronics with micromachining meets the need of low-cost and low-accuracy gyroscope that is applied in the car, control and consumer electronics. And the excellent properties of silicon micromechanical gyroscope determines that it has a wide range of commercial applications and military value, and thus countries think highly of the development of the gyroscopes.

The first silicon MEMS gyroscope was launched by America Draper Laboratory in 1991. It is a dual framework bulk silicon MEMS vibratory gyroscope, and the structure is shown in Figure 6.10. The plane size is 350 μm $\times$ 500 μm. Its random walk coefficient is $4\ °/(s \cdot Hz^{-\frac{1}{2}})$ when it is under the conditions of vacuum packaging and 60 Hz bandwidth. The gyro technology can be compatible with the microelectronic technology and also can be integrated with microelectronic interface circuit, so it is can be mass produced, and the cost is cut down subsequently[21]. Because of the dual framework silicon MEMS vibratory gyroscope was produced, a variety of silicon MEMS gyroscopes emerge endlessly.

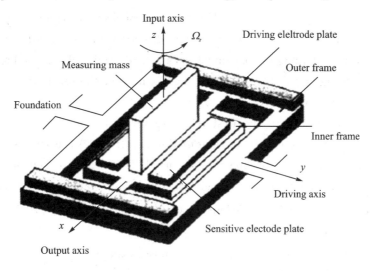

Figure 6.10 The dual framework gyroscope produced by Draper Laboratory

In 1993, Draper Laboratory invented a silicon MEMS vibrating tuning-fork gyroscope. The structure is shown in Figure 6.11[22-23]. The gyroscope was firstly introduced in the tangential comb drive in which its driving force is independent of the driving displacement. The electrostatic force drive those two sides of mass blocks opposite or away in the x-direction. When the y-direction angular velocity input, the Coriolis force leads to those two sides of mass blocks vibration up and down in the z-direction. The mass with the electrodes form a capacitance outputs the differential capacitance signal. The fabrication technique of the gyroscope includes reactive ion etching technology (RIE), silicon on glass bonding

Figure 6.11 The silicon micro vibration tuning-fork gyroscope developed by Draper

process, and dissolution technique. Working in the 60 Hz bandwidth, the random walk coefficient of the gyroscope is 1 $(°)/(s \cdot Hz^{-\frac{1}{2}})$, and the accuracy can meet the application of the low-accuracy military filed. In the drive direction of the comb drive tuning-fork gyroscope we will get a high Q factor, however, in the detecting direction due to the squeeze-film damping we will get a low Q factor. In order to improve the sensitivity of the gyroscope, it must be put in a vacuum cavity.

British Aerospace is one of the first teams to carry on the research on the ring mass gyroscope. They have applied it in automation control. As early as in 1994, the ring monocrystalline silicon gyroscope was reported. This kind of sensor is based on the glass substrates, and made by strong dry etching on the silicon chip of 100 μm, and when set in the air, the random walk coefficient is 0.05 $(°)/(s \cdot Hz^{-\frac{1}{2}})$, the scale factor is 20 mV/$[(°) \cdot s^{-1}]$.

In 1994, University of Michigan reported one vibrating ring gyroscope processed on the nickel plate. Its random walk coefficient is 0.1 $(°)/(s \cdot Hz^{-\frac{1}{2}})$, zero offset drift less than 10 $(°)/s$ (temperature range: $-40 \sim +85$ °C), the scale factor's nonlinear less than 0.2%(velocity range: ±100 $(°)/s$). As shown in Figure 6.12, the gyroscope is made up of one annulus, eight semicircle

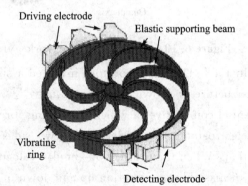

Figure 6.12 Vibrating ring MEMS gyroscope

elastic beams, and the drivers, the detectors, the capacitance electrode controllers evenly distributed around the annulus. The annulus will be driven by the electrostatic into the elliptic first bending mode processed the fixed amplitude in the plane. When the annulus is rotated around the normal axis, Coriolis force will cause the transfer of energy 45° offset from the first mode to the second mode. And the amplitude of the second mode is proportional to the angular velocity. The detecting electrode distributed around the annulus will detect the signal of the angular velocity. Compared with other gyroscopes, this gyroscope has the following characteristics: the symmetric structure that can reduce the interference from the environment vibration; the high sensitivity reached by matching the drive and detecting the modal resonant frequency; not sensitive to the change of temperature; and the production process error can be compensated by the electrode balance.

In 1995, Tanaka K.'s team from Japan developed the micromechanical gyroscope which works under a comb type electrostatic drive and the angular speed sensitive mass vibrates in the z-direction drive by Coriolis force, as shown in Figure 6.13[25]. The gyroscope is fabricated with polysilicon surface processing technology, and uses the ion etching technology to adjust structure parameters to match the resonance frequency of the driving-mode and detection-mode for high sensitivity. When the gyroscope works in the high vacuum under the 1 Pa pneumatic conditions, its equivalent noise angular velocity is 2 (°)/s. In the same year, Murata reported one axial (x or y axis) polysilicon surface gyroscope. It is developed by the phosphorus diffusion in the substrate and makes the detection electrode under the perforating polysilicon resonator. The random walk coefficient is 2 (°)/(s · $Hz^{-\frac{1}{2}}$).

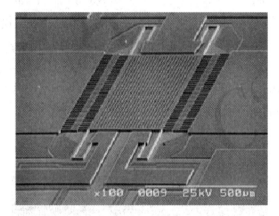

Figure 6.13 Micromechanical gyroscope vibrating in z-direction

In 1996, BSAC (Berkeley Sensor & Actuator Center) firstly reported one kind of MEMS gyroscope vibrating in the z-direction that fabricated with silicon surface microfabrication technology and with the structure of Comb Drive Actuator, comb capacitance detection, as shown in Figure 6.14. The gyroscope adopts the polysilicon surface MEMS processing technology, and the drive electrode and detection electrode are all comb finger structure. The frame comb interdigital constitutes the capacitance for detection and is driven by the vibration of the mass. The sensitive shaft is driven by Coriolis force vibrating perpendicular to the actuating shaft. The random walk coefficient is 1 (°)/(s · $Hz^{-\frac{1}{2}}$), the scale factor is 2 mV/[(°) · s^{-1}].

In 1997, BSAC had reported one xy biax gyroscope, as shown in Figure 6.15. The gyroscope is made on 2 μm thick polysilicon substrate with a disk rotor supported by elastic beam. The rotor is symmetrical in the x and y axes, thus it can be turned around two axes in

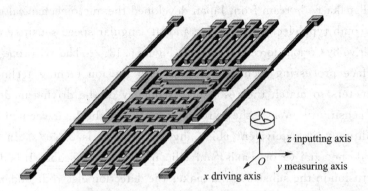

Figure 6.14 Vibratory MEMS gyroscope with comb driver and capacitance detector

the same conditions. When the gyroscope is working, the rotor will do torsional vibration around z axis forced by comb electrodes. When there is an angular velocity in x or y direction, the rotor will incline in y or x direction. And we get the angular velocity in x or y direction by detecting the capacitance change between the rotor and the substrate electrode. The random walk coefficient of the gyroscope is 0.24 $(°)/(s \cdot Hz^{-\frac{1}{2}})$, the scale factor is 2 mV/$[(°) \cdot s^{-1}]$, the lowest resolution is 10 $(°)$/h, and by the frequency matching, the resolution can be increased to 2 $(°)$/h.

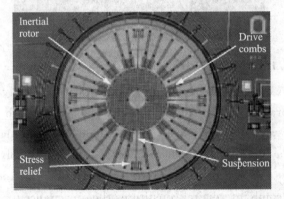

Figure 6.15 Biax MEMS gyroscope made by BSAC

In 1997, Samsung Company introduced one gyroscope made up with the polysilicon precipitation technology based on 7.5 μm thick low pressure chemical substrate. It gets a thin sensitive electrode only in 0.3 μm.

Also in 1997, with the help of the technology owned by British Aerospace Company, Robert Bosch Company designed a micromechanical tuning-fork gyroscope[30]. The company adopted electromagnetic prompting and capacitance detecting technology, and embedded the permanent magnet in the sensor, and these design techniques ensured some indicators. The amplitude of the actuating mode reaches, the bandwidth is 100 Hz, the random walk coefficient is 0.05 $(°)/(s \cdot Hz^{-\frac{1}{2}})$, the scale factor is 18 mV/$[(°) \cdot s^{-1}]$.

American JPL (Jet Propulsion Laboratory) cooperating with UCLA (University of California, Los Angeles), introduced one alfalfa leaf bulk micromachining gyroscope, as shown in Figure 6.16. The structure is four vanes braced by four slender beams, and in the middle there is one metal bar assembled. The vanes D1 and D2 are the electrostatic driving electrode and drive the vanes twist around y axis. When the angular velocity around z axis input, the Coriolis force will couple to the torsional detection mode in x direction from the torsional mode in y direction, by detecting the capacitance change of the detection electrode S1 and S2 to obtain the torsional amplitude. The manufacturing process is very complex and needs to ensure the central points of the mental bar and the four vanes perfect alignment. So the process is very difficult. And in order to increase the rotational inertia force in detection, we need to inject metal epoxy into the holes of the silicon resonator. The random walk coefficient is 90 $(°)/(hr \cdot Hz^{-\frac{1}{2}})$, the scale factor is 24 mV/$[(°) \cdot s^{-1}]$.

Figure 6.16 MEMS gyroscope in alfalfa leaf type

In 2000, Mochida Y. et al. from Japan proposed one MEMS gyroscope with the independent drive beam and detection beam structure, and did the research on the coupling problem between different modes[32]. The structure schematic diagram is shown in Figure 6.17. The outside beam is the driven beam, the mass and frame will vibrate in the driven direction forced by the electrostatic force. When the angular signal is input, due to the effect of the Coriolis force, the mass vibrates in the detecting direction. The inner beam either as the detection beam, or isolates the coupling between detection mode and driven mode. This helps to reduce the quadrature error, and improve the gyroscope performance. And in the vacuum environment (less than 100 Pa), the quality factor of the detection direction is 1 000, and working at 10 Hz bandwidth, the equivalent noise angular velocity of the gyroscope is 0.07 (°)/s.

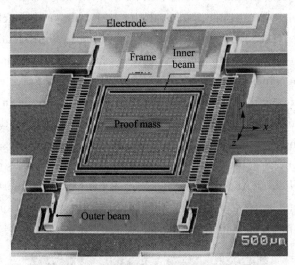

Figure 6.17 Tuning-fork MEMS gyroscope with the independent beam

In 2001, Carnegie-Mellon University reported one gyroscope which got the xy axes gyroscope[33] and z axis gyroscope[34] integrated. Its random walk coefficient is 0.5 (°)/(s · $Hz^{-\frac{1}{2}}$). The technologies of COMS (no mask) and micro machined process are adopted to fabricate the gyroscope. The xy axes gyroscope makes 0.5 μm COMS metal thin film layer on a 5 μm substrate. And Z axis gyroscope makes 0.18 μm COMS metal thin film layer on an 8 μm substrate.

In 2002, HSG-IMIT reported a silicon micro machined vibratory wheel rate gyroscope, as shown in Figure 6.18. This kind of structure has excellent decoupling ability in driving mode and detection mode. It is made up on 10 μm polysilicon substrate with the standard Bosch technology. When working at 10 Hz, the equivalent noise angular velocity is 0.025 (°)/s, the bias instability is ±0.3 (°)/s, the random walk coefficient is 25 (°)/(h · $Hz^{-\frac{1}{2}}$), the scale factor is 10 mV/[(°)/s^{-1}].

In 2002, BSAC reported an integrated z axis MEMS gyroscope which output the frequency detection directly. The structure is shown in Figure 6.19. This kind of gyroscope and the gyroscope reported in 2000 both use the MEMS technology which is compatible with

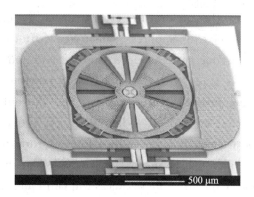

Figure 6.18 Vibratory wheel rate gyroscope made by German HSG-IMIT

COMS, and the MEMS technology made in National Laboratory. But the improvement is that the detection chips and detection circuit are integrated. This helps to increase the reliability. At the atmospheric pressure, the random walk coefficient is $0.3\ (°)/(s \cdot Hz^{-\frac{1}{2}})$, which is the same as that of the gyroscope reported in 2000.

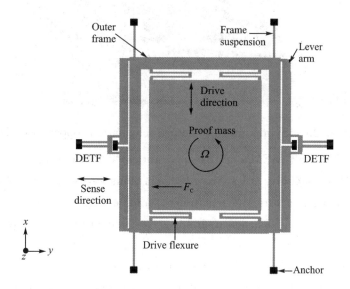

Figure 6.19 Directly output frequency resonant gyroscope made by BSAC

In 2002, Alper S. E. et al. in Middle East Technical University proposed a new kind of symmetrical MEMS gyroscope[37] structure (shown in Figure 6.20).

The gyroscope can not only achieve the frequencies matching to the driving mode and the detection mode to obtain a high sensitivity, but also can decouple the driving mode and the sensitive mode to eliminate the quadrature error caused by cross-coupling. The fully symmetric structure is conducive to suppress sensitivity drift caused by temperature and process technique. The gyroscope gets the electrostatic drives and comb capacitive detection mode, the interface circuit is made with 0.8 μm CMOS fabrication technique. Alper etc. used surface micromachining process, LIGA process and the bulk silicon micromachining technology to fabricate the structure. When the gyroscope made by the surface micromachining fabrication technology is under 10 mTorr pressure conditions, the quality factor will be 10 400, and the resolution will be up to 1.6 (°)/s. When the gyroscope is made with LIGA fabrication technology, the structure thickness will be 16 μm. The random walk coefficient in vacuum conditions is 7.3 (°)/(s·$Hz^{-\frac{1}{2}}$). When working at 50 Hz bandwidth, resolution will be 96(°)/s. Because the structure gets the static capacitance in the detection direction, it processes the drawbacks of small sensitivity.

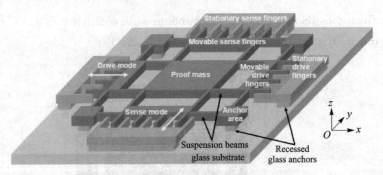

Figure 6.20 Symmetrical structure schematic proposed by METU

In 2005, a kind of MEMS gyroscope fabricated with single crystal silicon was reported[38]. Its bias instability is less than 1 (°)/h. This gyroscope works in mode matching area, so it has high scale factors. Its sensitivity is a little low. It fabricated with the common bulk silicon technology, so it has a coarser surface. Its superior performance is mainly due to the 10 mTorr vacuum level.

Since 2008, the detection principle and structure of gyroscope have come into stereo type state. The development focus has shifted from structure innovation to the performance optimization, mainly in the following trends:

(1) In order to reduce the coupling, many kinds of gyroscopes are designed in multiple masses and using the difference principle, as shown in Figure 6.21. As for a double degree of freedom gyroscope, the focus of the design work is on limiting the amplitude of the mass in the driving detection or the detection direction. When the Coriolis force generated by the

driving mass imposes on the detection mass, it will restrict the detection mass vibrating in the driving direction, even lead to the driving mass can't vibrate in the detection direction.

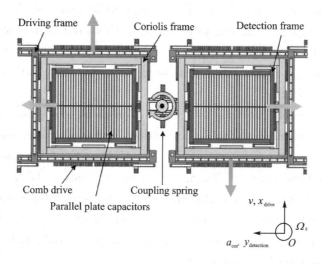

Figure 6.21 Double mass symmetrical decoupling gyroscope

(2) When it is not subjected by the external forces, we call it free vibration; when subjected to external forces, we call it forced vibration. The gyroscope drive is actually a periodic forced vibration. If the driving frequency of the external force and the detection frequency are as close as possible to each other, which is called resonance, the system will harmonic vibrate with the maximum amplitude, but also obtain maximum sensitivity. Of course, this is an ideal state. In recent years many studies have tried to optimize the matching of these two frequencies, but require sophisticated processing methods to match the two frequencies of the driving mode and the detection mode.

(3) Because a large part of the gyroscope performance depends on the structure of production, even if the performance is well-designed, but due to the limitation from micro size constraint and processing technology, there are some inevitable flaws. Because of these manufacturing defects, in the detection signal there may be a large error, which requires a series of compensation methods. At this stage there are a lot of researchers that have attempted to apply the control concept to the driving and detection circuit. Literature[39] reported a practical correlation filter can effectively reduce the thermal noise compared to the conventional low-pass filter. Literature[40] reported one method to reduce noise in use of wavelet analysis. Leland et al. reported an adaptive controller [41] that is possible to estimate the coupling effects and compensate the influence of the effect. Zheng et al. reported a new

type of oscillation control[60] in the driving direction. The controller adds a linear state observer to the traditional PD controller, and it can effectively inhibit the uncertainties from the structure of the actuating shaft. The design philosophy of the controller gets the following advantages: set the driving or detection methods in the state in our needs; enhance the dynamic response time and enlarge the dynamic input range; estimate the unknown parameters in real-time; compensation processing defects; compensation time-variant effects such as temperature effects.

(4) Another research direction is a combination of materials and chemistry to explore new materials that are more suitable than silicon, and new processing methods that can better improve gyroscope performance. National University of Singapore proposed a new material called graphene, and its Young's modulus reaches 1 TPa, and tensile strength reaches 130 GPa, which are much higher than the Young's modulus of silicon (165 GPa) and tensile strength (5~9 GPa). According to the equation of frequency, it is easy to draw a conclusion that the graphene gets a higher resonant frequency and reliability than silicon, but these are still in theoretical assumption state.

6.2.3 Classification of micromechanical gyroscope

There are many types of micromechanical gyroscope, in order to facilitate the analysis, the gyroscopes are classified according to materials, vibration structures, driving or detection methods and working modes.

(1) According to different vibrating structures, the micromechanical gyroscopes can be classified into linear vibration structure and angular vibration structure. The linear vibrating structure can be further classified into orthogonal vibrational structure and non-orthogonal vibrational structure.

(2) According to different materials, the gyroscopes can be divided into silicon gyroscope and the gyroscope not in silicon material. The silicon material includes the monocrystalline silicon and the polycrystalline silicon; other materials are mainly for piezoelectric quartz and piezoelectric ceramics.

(3) According to different processing methods, the gyroscopes can be divided into LIGA gyroscope, bulk micromachining gyroscope and surface micromachining gyroscope.

(4) According to different driving methods, the micromechanical gyroscopes can be divided into electrostatic gyroscope, electromagnetic gyroscope and piezoelectric gyroscope etc.

(5) According to different detection methods, the gyroscopes can be divided into

capacitive gyroscope, resonant gyroscope, piezoresistive gyroscope, piezoelectric gyroscope, optical gyroscope and tunneling gyroscope, which are concluded in Figure 6.22.

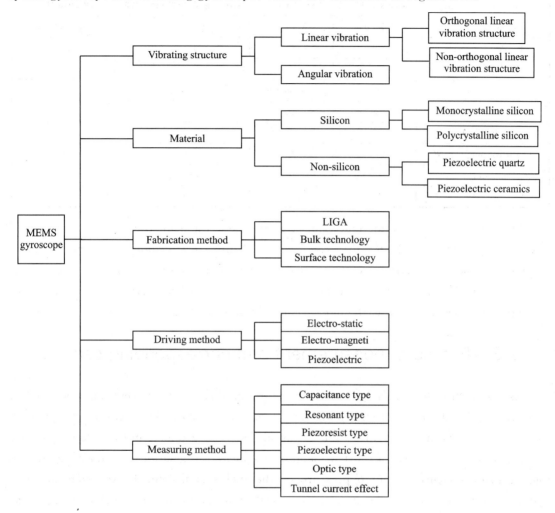

Figure 6.22 The classification of MEMS gyroscopes

According to different materials, the gyroscopes can be divided into silicon gyroscope, quartz gyroscope and piezoelectric ceramic gyroscope. From the 1980s, the most representative gyroscopes are the silicon micromechanical gyroscopes developed by American Draper Laboratory, the quartz micromechanical gyroscopes developed by Siston Donner Company and the piezoelectric ceramic micromechanical gyroscopes developed by Japanese Murata Company. The parameters of the micromechanical gyroscopes are listed in Table 6.1.

Table 6.1 Performance parameter of MEMS gyroscopes[19]

Gyroscope model	Gyrostar ENC	Gyrostar ENV	Gyrochip GC – 1 – 50 – 100	ВОГ910 НТК
Company name	Murata	Murata	Siston Donner	ФИЗОПТИКА
Range/[(°)·s^{-1}]	90	80	100	200
Frenquency/Hz	50	7	60	1000
Drift rate/[(°)·s^{-1}]	—	1(10 min)	0.2	0.01
Nonlinear/%	5	0.5	0.1	0.2
Sensitivity Axis	1	1	1	1
Power consumption/W	0.025	0.075	0.8	0.5~1.5
Dimension/mm	22×9×8	19×46×38	Φ50×25	Φ80×25
Mass/g	2.7	50	100	100

The research of silicon micromechanical gyroscope is the mainstream. From the view of driving and detection methods, they can be divided into electromagnetic driving capacitive detection gyroscope, electromagnetic driving-resistance detection gyroscope and electrostatic driving-capacitance detection gyroscope. For most of the silicon gyroscope, electrostatic driving-capacitance detection is the most commonly used method.

6.3 Pressure measurement and corresponding sensors

As one of the atmospheric parameters, the pressure parameter is very important and its measurement has been a crucial testing direction. The sensors that measure the pressure parameters have been applied in various technical fields, for example in the field of aviation. The pressure in import and export of aviation engines, the aviation air-data computers in flight control system, oil pump pressure of the fuel system, the brake arrangement in hydraulic system, air conditioning system, as well as navigation and fire control systems, all of the above systems require the pressure sensor to provide relevant pressure parameters. And with the development of micromechanical technology, the main development directions of the micromechanical gyroscope are small size, low power consumption, high integration, low cost, and high performance.

6.3.1 Working princinple

Currently, the vast majority of the sensitive elements in silicon MEMS pressure sensors are made up with silicon diaphragm, and there are three sensitive mechanisms: the piezoresistive diaphragm based on the physical effects, the capacitive diaphragm based on the laws of physics and the mechanical resonant diaphragm. Figure 6.23 is the working principle schematic of the three sensors (corresponding diaphragm theory refers to Chapter 2). Figure 6.23 (a) shows a silicon piezoresistive pressure sensor, the piezoresistor diffused on the silicon diaphragm and formed a Wheatstone bridge; when the measured pressure presses on the diaphragm, the diaphragm will be deformed causing changes in the resistance of the piezoresistor, and the bridge becomes imbalance, the loss is proportional to the measured pressure. This kind of pressure sensor assembly gets the advantages of easy produced and assumed, and high yield etc. But, because it is based on the piezoresistive effect of semiconductor, it is subjected to temperature effect. It is difficult to achieve high-precision measurement of pressure. Figure 6.23 (b) shows a silicon capacitive pressure sensor: the metal layer deposited on the lower diaphragm makes up the movable electrodes of the capacitance, the other electrode is deposited on the substrate surface of the silicon, and the two electrodes constitute the parallel-plate capacitor. When the diaphragm bending occurs caused by the pressure, the capacitor plate spacing changes, causing the change in capacitance and the change corresponding to the measured pressure. This kind of pressure sensors won't be affected by the temperature changes, so the measurement accuracy won't be affected by the temperature. Meanwhile, because the capacitance detection is based on the detection of analog implementations, the overall detection accuracy is related to the weak capacitance detection accuracy. Meanwhile, converting the digital/analog signal will cause the large test circuit, and will affect measurement accuracy. Currently, the most high-profile sensor is the third pressure sensor, namely silicon resonant pressure sensor, which is based on the resonant frequency of the diaphragm or the beam changes corresponding to the measurement pressure to achieve the pressure measurement. The typical structure is shown in Figure 6.23(c). Silicon diaphragm (or beam) can be driven by electrostatic or other motivation source to generate harmonic vibration, the resonant frequency is f_o. When the diaphragm (or beam) is subjected to the measured pressure directly (or indirect), the stiffness will change, resulting in change in the resonant frequency, which is corresponding to the measured pressure, then obtain the corresponding pressure value by calculating, because the sensor is based on the frequency detection to achieve the measurement of the pressure, which has some primary attributes: good stability, digital output and good precision.

In the three silicon micromechanical pressure sensors, the silicon piezoresistive pressure sensor is the first developed. Back to the 1950s, the USA Bell Laboratories discovered

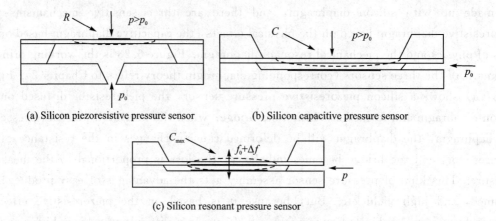

Figure 6.23 Three typical silicon MEMS sensors

silicon piezoresistive effect [42]—the strain coefficient is twice higher than the strain coefficient of metallic materials. And in the early 1960s, the bulk silicon strain gages were glued on the metal diaphragm directly to replace the metallic stain gages, and then made up the silicon diaphragm to replace the metal diaphragm and diffused silicon piezoresistive on which formed the current silicon piezoresistive pressure sensors, this helped significantly improve the performance of the sensor, but also reduced the production cost. In the 1970s and 1980s, with the further development of silicon micromachining technology, especially the development and application of silicon surface processing technology help to get better performance of micro and ultra-micro silicon piezoresistive pressure sensors. They got smaller size (e.g. the diaphragm size is smaller than 1 mm×1 mm, the thickness is less than 5 μm), lower power consumption, and better suited to the needs of the aerospace and biomedical fields, and easier monolithic integrated with microelectronic circuits. Due to its successful application and wide range of areas, it becomes a world well-known pressure sensor. But the main drawback of this kind of sensor is temperature instability, and when it is applied in the large temperature changeable ambient, the temperature compensation is necessary. In addition, limited by the piezoresistor sensitivity, the sensor doesn't apply to the accurate measurement of ultra-low dropout.

The conventional capacitive pressure sensor which was made up with metal or ceramic components had been applied in many industrial fields, for example back to 1940s the capacitive fuel gauge had been used to detect the oil level on the plane. From the late 1960s to the 1980s, the capacitive differential pressure sensors and transmitters are widely used in industrial process monitoring, and the products produced by the United States Rosemount

Company and Japan Yokogama Electric Company are representative and dominant. After entering the 1980s, with the development and mature silicon MEMS and integrated circuit technology, the integrated silicon capacitive pressure sensor with the silicon diaphragm as the active plate obtained a rapid development and wide application. Compared with silicon piezoresistive pressure sensor, the advantages of silicon capacitive pressure sensor that makes it outstandes, mainly includes high sensitivity, large measuring range from the SPL to tens of MPa; lower power consumption (two orders of magnitude less than piezoresistive consumption); and the long-term stability and repetition are better than silicon piezoresistive. In view of this, the silicon capacitive pressure sensors are well-known, which are widely used in airborne atmosphere data measurement and control systems, industrial process control, clinical monitor, and biomedical research etc. But there are also some disadvantages in silicon capacitive pressure sensors, especially in the micro-pressure measurement. The change of the micro-capacitance is susceptible to the interference of stray capacitance, so its interface circuits are much more complex than silicon piezoresistive pressure sensor. In order to avoid the interference of stray capacitance, the interface circuit is preferably integrated on the sensor chip, or designed proximity to the sensor.

6.3.2 Resonant silicon micromechanical pressure sensor and its development

Compared with the piezoresistive and capacitive micromechanical silicon pressure sensors, resonant silicon micromechanical pressure sensor obtains the measured pressure by detecting the resonant frequency change of the sensitive element before and after the load, therefore it has the advantages of good stability, digital output, high precision etc., and become the mainstream of high precision pressure sensors.

The research of resonant pressure sensor had begun early in 1919. The vibrating wire pressure sensor appeared earlier, and is still in use on some occasions. Subsequently, until 1959 there came the second generation of resonant pressure sensor, namely the resonant cylinder pressure sensor renowned in aviation. Since 1960, the branch company (former Solartron Company) of the United Kingdom Schlumberger Industries has been developing the resonant cylinder pressure sensor. Since the 1970s of the 20th century, the sensor has been widely used in aero engine control system and the air data computer system, and becomes the high-end technology in the Airborne System.

In the 1980s, with the rapid development of silicon MEMS technology, these resonant sensors based on the precision machining started porting onto a silicon substrate, which means that the silicon micro-machining technology was applied to produce silicon resonant pressure sensor. Since the resonance diaphragm directly contacts with the gas to be measured, it will cause vibration energy of the micro-diaphragm dissipation, leading Q value too low to start-up; in addition, because of the measured gas acting directly on the resonant diaphragm, the resonant frequency is not only dependent on the pressure, but also dependent on the mass, type, temperature of the gas near the diaphragm, and the absorption of chemical or dust. The corrosive effects will change the mass of the resonator, cause the resonant frequency drift, and reduce the accuracy of the sensor. Therefore, this silicon resonant diaphragm pressure sensor is not practical.

Because of the above shortcomings, British scientist Greenwood J. C. did a pioneering research on silicon resonant pressure sensor. He proposed one new silicon resonant pressure sensor adopted the composite sensitive structure, and published in 1984[43], This paper is a milestone foreboding the silicon resonant pressure sensor begin to be practical. The structure is formed by anisotropic etching on silicon substrate, including square silicon diaphragm and resonant beam supported on the surface of the diaphragm at the center, and both constitute a single entity as the chip in silicon resonant pressure sensor (shown in Figure 6.24). Figure 6.25 shows an electron micrograph of the resonant beam which shapes a butterfly, so it is called butterfly beam.

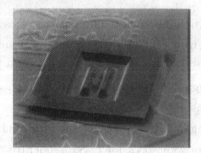 

Figure 6.24 Chip in silicon resonant pressure sensor

Figure 6.25 Electron micrograph of the resonant beam

Because the resonator is isolated with the measured medium, this kind of silicon resonant pressure sensor based on sensitive composite structure shows superior performance in the gas and liquid pressure measurement, and it is conducive to industrialization. The sensor has been highly recognized by the international peer. Now, several well-known international companies are developing and manufacturing distinctive resonant silicon pressure sensor based on this structure. In 1988, Greenwood J. C. published his findings[44] again, and the paper was mainly about the content enriching of the paper published in 1988 and some corresponding theoretical analysis. The design was adopted and improved by British Druck Inc. (now it is part of GE), and ultimately achieved industrialization. In the late 1990s, this kind of sensor was rolled out to the consumers.

As shown in Figure 6.26, silicon beam supported on the ribbed silicon diaphragm, both constitute a single entity. The silicon resonant beams get 6 μm in thickness and 600 μm in length, driven by electrostatic actuation, and adopt the capacitance detection. The glass tubes in the cavity achieve vacuum packaging. In the 100 kPa pressure range, the resonant frequency change rate is about 9%; full scale accuracy is better than 0.01%; annual stability is better than 0.01%. Figure 6.26 shows the sensor measurement result of the pressure frequency characteristic. The sensor has been successfully applied as weather gauge, precision pressure calibration equipment, and becomes one of the world-renowned silicon resonant pressure sensor products.

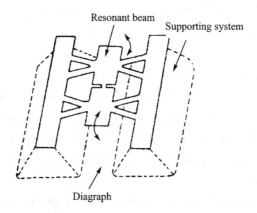

Figure 6.26 Sketch of the sensitive structure

Schlumberger's aviation sensor branches reported one kind of silicon resonant pressure sensor in the literature[45-46] published in 1991 to 1992, which was driven by electrostatic and detected by piezoresistive effect, and met the needs of advanced airborne electronic equipment. And in the chip, the upper and lower silicon slices constituted a single entity by using the Si-Si bonding technology, and the double-clamped beams were set on the center shallow groove on the upper surface of the diaphragm, as shown in Figure 6.27. The diaphragm size is 2.5 mm × 2.5 mm; the resonant beam size is 600 μm × 40 μm × 6 μm; the silicon resonant fundamental frequency is 120 kHz. The Q value of the silicon beam is 60 000 under vacuum packaged, and the Q value of some sensors that are whole packaged may be up to 140 000. This means that the width is only 1 Hz at the resonance peak of 120 kHz, so the resolution is easy to reach the order of 1 ppm.

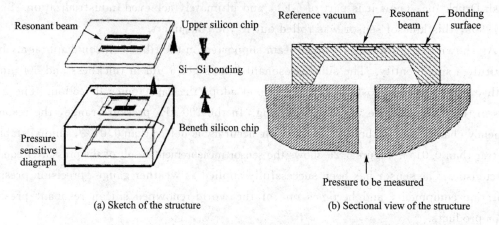

Figure 6.27 Sensitive structure of the resonant pressure sensor

In the mid-1990s, Japan Yokogama Company developed a silicon resonant differential pressure sensor, its sensitive structure is shown in Figure 6.28[47-48]. The pressure sensing diaphragm is firstly fabricated on the silicon chip whose size is 6.8 mm×6.8 mm×0.5 mm. Then, two H-type resonant beams (1 200 μm×20 μm×5 μm) is fabricated on the surface by using the Boriding extension technology, and one beam is set at the center of the diaphragm, while the other is set at the edge of the diaphragm, so as to form a complex feeling pressure. Silicon beam is encapsulated in a partial vacuum chamber by reaction sealing techniques, and the Q value is up to 50 000.

When the positive pressure acts on the diaphragm, the bending deformation will take place. When two resonant beams are forced by opposite directions stress, the resonant frequency of the center beam will increase because of the tension. Meanwhile, the frequency of the beam at the edge will reduce because of the compression. The frequency difference between the two resonant beams corresponds with the difference of the measured pressure. The method by testing the frequency difference to obtain the measured pressure can not only improve the sensitivity of the measurement, but also have strong inhibitors on the common mode interference such as the ambient temperature.

Druck Company, Yokogama Corporation and Schlumberger Company are the world's most prestigious manufacturers in silicon resonant pressure sensor production. The productions of Druck Company and Yokogama Corporation are mainly for civilian service,

Chapter 6 MEMS testing technology and engineering application

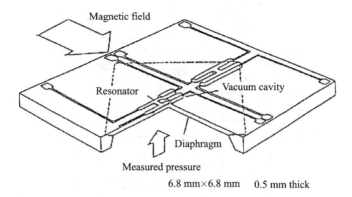

Figure 6.28 Sensitive structure of the pressure sensor

while Schlumberger Company focuses on developing high precision silicon resonant pressure sensor for aviation. The technology is mature enough in UK, USA, Japan, Netherlands and other developed countries to be applied in many precision measurement fields.

In China, since the mid-1980s, Beihang University had begun to research and develop the silicon resonant pressure sensor on the background of a new generation micro-sensors for aviation, and founded the Micro-sensor Technology Laboratory. And in the early 1990s, Beihang University first proposed a scheme about a silicon resonant beam pressure sensor that is driven by electro thermal and piezoresistive detection. This scheme was funded by AVIC during the "95", subsequently cooperating with the relevant MEMS Institute and aviation cooperation. Using Si—Si bonding technology and SOI technology, Beihang University manufactured a small batches of experimental samples, and the sensitive structure is shown in Figure 6.29. And we independently developed a dedicated tester for the sensor to obtain the open-loop frequency characteristics (including the detection acquisition, processing of weak vibration signal, and the display of testing control and testing results). Then a lot of test experimental and analytical studies had been done on the samples, and a number of key technologies had been mastered, and some promising initial results had been made.

Through experiments we also found a number of problems, including:

(1) The static temperature rise of the electro thermal leads to the resonant beam generating the axial compressive stress, and leads to the resonant frequency of the beam drifting. In addition the changes in ambient temperature will also generate the significant impact on the resonant frequency of the beam. Therefore the thermal stability becomes an important factor in limiting of the sensor high-performance, so it is necessary to optimize the design and do some proper temperature compensation. This is just the main content of this

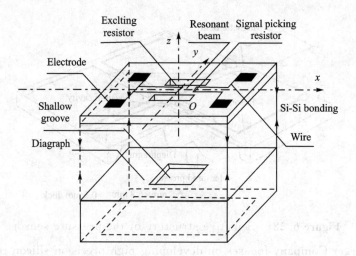

Figure 6.29 Pressure sensor with electro thermal and piezoresistive detection subject.

(2) Because the silicon micromachining, micro-assembly and packaging technologies in domestic, there is a wide gap among the developed countries, so far we have never manufactured an engineering prototype of the silicon resonant beam pressure sensor with high Q value, stable and reliable. Currently, the Q value of the sample is only thousands, compared with the international level. The difference is $1 \sim 2$ orders of magnitude. Therefore, it is difficult to achieve the closed-loop self-excited oscillation.

6.4 Measurement of micro-torque

6.4.1 Introduction

MEMS is a very promising interdisciplinary field that developed from microelectronics in recent years. MEMS has wide applications in many fields such as medicine, military, aeronautics and astronautics, micro-robot etc. Being one of the major power sources of MEMS, micro-motors are always the major research direction for researchers. But, with the development of MEMS theory and fabrication technology, the dimension of micro-motors becomes smaller and smaller and the output micro-torque decreases correspondingly[49-51]. How to exactly measure the output micro-torque becomes an important research issue. One of the traditional measuring methods is called contact method. Mathiesont D. et al.[52] use the viscous braking method to measure the micro-torque of the motor. The tested precision is about 10^{-5} N·m. Gass V. et al. use the so-called cantilever method to measure the output micro-torque[53]. The tested resolution of the device is 0.05 μN·m. Brenner W. et al.[54] use cable brake principle for the measurement of torque-characteristic of mini-motors and micro-motors. The tested resolution is about 1 μN·m and the relative error is about 7.5%. Although many effective results have been obtained by this method, the micro-torque is so little (μN·m order) that many outer factors caused by the contact method such as measuring temperature, vibration of the equipment, frictional force and flow of the air, may have a great influence to the tested results. It becomes more difficult to obtain high precise result so as the device under measurement becomes smaller, for there is a considerable increase in the loss caused by the mechanical contact between the motor and the measuring equipment. Another kind of method is the non-contact method, in which there is no mechanical contact between the micro-motor and the testing equipment. The interaction between them can be initiated by means of magnetic induction between two cores [55] or by air pressure [56-57].

A newly-developed noncontact measuring method has been introduced to measure the 10^{-6} N·m order micro-torques. It uses electromagnetic force to apply a load to the micro-motor and the braking force is controlled easily by changing the DC braking voltage applied to the windings. In order to avoid the measuring errors caused by the theoretical calculation, an electronic balance is used to calibrate the measured braking force. Because the micro-motor and the measuring device is detached from each other during the experiment, outer factors caused by the contacting measuring method may have very small effect to the micro-motor. Compared to the measuring method mentioned before, the braking force can be directly and timely obtained by a high precise electronic balance. This avoids the measuring errors caused by the conversion of the force and theoretical calculation. Frictional force and vibration etc. caused by the contact method are also avoided. The total measuring precision is increased correspondingly.

6.4.2 Working principle of noncontact method

Simple working principle of noncontact method is shown in Figure 6.30. The measuring system mainly consists of a malleable cast iron core with windings all around it and an aluminium disk that is fixed on the output axis of a micro-motor. In the middle part of the magnetic core there is a small gap (about 3 mm) in length (shown in Figure 6.31).

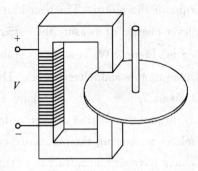 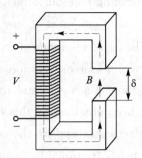

Figure 6.30 Working principle of the noncontact method

Figure 6.31 Calculation model of the magnetic induction intensity

Border of the aluminium disk is placed in the gap. During the course of the experiment a certain driving voltage is firstly applied on the micro-motor and the aluminium disk begins to rotate together with the output axis of the micro-motor. Then, a DC braking voltage is applied on the windings and the electromagnetic field is filled in the gap. According to electromagnteic theory, when the aluminium disk is rotating perpendicular to the direction of the line of magnetic force, the electromagnetic force, which is also the braking force, is applied on the border of the aluminium disk. The product of the braking force and the radius of the aluminium disk is the output torque of the micro-motor. Rotating speed corresponding to the output torque can also be measured by the speed measuring equipment.

6.4.3 Theoretical calculation

6.4.3.1 Calculation of the magnetic induction intensity in the air gap

Figure 6.31 is the simple calculation model of the magnetic induction intensity. This model is deduced on the assumptions as follows:

(1) The magnetic line of force in the air gap is parallel to each other and the magnetic field's edge effect is ignored.

(2) The magnetic permeability of the malleable cast iron is so big (μ_I, which represents the initial permeability, is about 10 000; μ_M, which represents the maximum permeability, is about 200 000) that it centralizes all the magnetic flux inside it. No magnetic flux is leaked to the surrounding air.

(3) Electric resistance of the aluminium can be divided into two parallel connected portions—the one in the air gap (which is facing the magnetic pole) and the lefted protion.

(4) Aluminium material which is placed in the gap of the rectangle magnetic pole can be regarded as countless parallel arranged aluminium wire. Every wire has the same width, height and length. So, their electric resistances are the same.

According to the Ampere's theorem of electrocircuit when a certain DC driving voltage is applied on the windings of the magnetic pole, corresponding electric current I in the windings can be expressed as

$$I = \frac{V}{R_w} \tag{6.8}$$

where V is the DC driving voltage applied on the windings, R_w is the electric resistance of the windings.

According to the assumptions and Ampere's circuital theorem in the magnetic field, the relationship between I and the magnetic induction intensity B in the air gap (the simplified calculating model is shown in Figure 6.31) can be expressed as

$$NI = \frac{\varphi l}{S_1 \mu_1 \mu_0} + \frac{\varphi \delta}{S_2 \mu_0} \tag{6.9}$$

where N is the turn of the windings around the malleable cast iron core; φ is the magnetic flux; l is the length of the malleable cast iron core along the magnetic circuit; δ is the length of the air gap; μ_1 is the relative magnetic permeability of the material; μ_0 is a constant, $\mu_0 = 4\pi \times 10^{-7}$; S_1 and S_2 are areas of the cross section in the ferromagnetic core and in the air gap. In this design, $S_1 \approx S_2$, so $\frac{\varphi}{S_1} = \frac{\varphi}{S_2} = B$, in which B represents the magnetic induction intensity in the magnetic circuit. Equation (6.9) can be substituted as

$$NI = \frac{B}{\mu_0}\left(\frac{l}{\mu_1} + \delta\right) \tag{6.10}$$

where N, l, μ_0, μ_1 and δ are all known. When I is applied on the windings, B can be calculated correspondingly, i.e.

$$B = \frac{\mu_0 \mu_1 N}{l + \mu_1 \delta} I \tag{6.11}$$

The magnetic flux Φ in the air gap can be expressed as

$$\Phi = \iint_S \vec{B} \cdot d\vec{s} = BS = \frac{\mu_0 \mu_1 N}{l + \mu_1 \delta} ab I \tag{6.12}$$

6.4.3.2 Calculation of the braking micro-torque

Simple position relationship between the aluminium disk and the braking magnetic pole is shown in Figure 6.32. In this figure, symbol $\otimes$ represents the direction of magnetic field.

According to assumption (1), on the aluminium disk, the magnetic field just has effect on the rectangle area that is facing the rectangle braking magnetic pole. According to assumptions (3) and (4) in Subsection 6.4.3.1, the aluminium in the air gap of the magnetic pole can be divided into countless parallel connected slim wires. The width of each wire is supposed to be the infinitesimal value dy and the length is a. Each wire has the same resistance, which is $r_1 = r_2 = \cdots = r_n = r_{equal}$. The connecting relationship between the portion facing the braking magnetic pole and the other portion on the aluminium disk is shown in Figure 6.33. Simple calculating model of the induced electric potential for each wire is shown in Figure 6.34. In this model the origin of the calculating coordinate system is placed at the center of the aluminium disk. Supposing there is a little element of which the length is dx and the coordinate is (x, y) $(a \leqslant x \leqslant b)$ on the wire. According to the theorem on electromagnetics, when the aluminium is rotating around the origin, the induced electric potential for each little element dx can be expressed as

$$d\varepsilon = (\vec{v} \times \vec{B}) \cdot d\vec{x} = Bv\cos\theta dx \tag{6.13}$$

where v is the moving linear velocity of dx and θ is the included angle between dx and v.

$$\cos\theta = \frac{x}{r} \tag{6.14}$$

where r is the rotating radius of dx, and

$$r = \sqrt{x^2 + y^2} \tag{6.15}$$

v in equation (6.13) can be expressed as

$$v = \omega r = 2\pi f r = 2\pi \cdot \frac{n}{60} \cdot r = \frac{\pi n r}{30} \tag{6.16}$$

where n represents the rotating speed of the aluminium disk, and the unit of it is rpm.

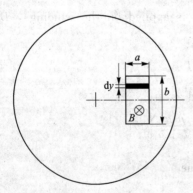

Figure 6.32 Position relationship between the aluminium disk and the braking magnetic pole

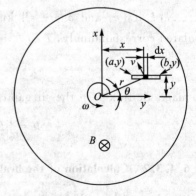

Figure 6.33 Calculation model of the induced electric potential

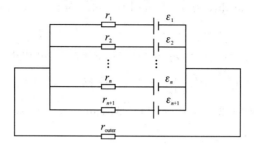

Figure 6.34 Connecting relationship among the portions on the aluminium disk

Thus when dx is rotating around the axis, the corresponding induced electric potential according to equation (6.13) can be expressed as

$$d\varepsilon = B \cdot \frac{n\pi}{30} \cdot \sqrt{x^2 + y^2} \cdot \frac{x}{\sqrt{x^2 + y^2}} \cdot dx = \frac{n\pi B}{30} \cdot x dx \quad (6.17)$$

From equation (6.17) we can see that in the defined gap area, every element whose x is equal to each other initiate the same induced potential. We call the cross section with a same x coordinate equipotential surface, so the total induced potential along the wire can be expressed as

$$\varepsilon = \int_a^b d\varepsilon = \frac{n\pi B}{30} \int_a^b x dx = \frac{n\pi B}{60}(b^2 - a^2) \quad (6.18)$$

Because a and b are constants in equation (6.18), we get

$$k = \frac{\pi}{60}(b^2 - a^2) \quad (6.19)$$

Then equation (6.18) can be substituted as

$$\varepsilon = knB \quad (6.20)$$

According to Figure 6.34, because all the induced potentials in the circuit are parallel connected, the total induced potential in the circuit is also ε. The total resistance can be expressed as

$$R_{rot} = r_{outer} + \frac{r_{equ}}{n} \quad (6.21)$$

so the eddy current in the circuit can be expressed as

$$I_{rot} = \frac{\varepsilon}{R_{rot}} = \frac{knB}{R_{rot}} \quad (6.22)$$

According to Ampere's equation in electromagnetics, when I_{rot} is flowing along the aluminium disk, the Ampere's force can be expressed as

$$F = \int_a^b dF = \int_a^b I_{rot} dx \times B = BI_{rot}(a - b) \quad (6.23)$$

Micro-torque in vertical direction of the little element can be expressed as
$$dM = I_{rot}\,d\vec{x} \times \vec{B} \cdot x = I_{rot} Bx\,dx \tag{6.24}$$

The total vertical micro-torque applied on the aluminium disk can be expressed as
$$M = BI_{rot} \int_a^b x \cdot dx = \frac{1}{2}(b^2 - a^2)BI_{rot} \tag{6.25}$$

Combining equations (6.22) and (6.25), M can be expressed as
$$M = \frac{1}{2}(b^2 - a^2)BI_{rot} =$$
$$\frac{1}{2}(b^2 - a^2)B\frac{knB}{R_{rot}} = \frac{knB^2(b^2 - a^2)}{2R_{rot}} \tag{6.26}$$

Because k, a, b and R_{rot} in equation (6.26) are all constants, and take $m = \frac{k(b^2 - a^2)}{2R_{rot}}$, then equation (6.26) can be substituted as
$$M = mnB^2 \tag{6.27}$$

Combining equations (6.8), (6.11) and (6.27), the relationship between the braking voltage and the micro-torque can be expressed as
$$M = mnB^2 = mn\left(\frac{\mu_0 \mu_1}{l + \mu_1 \delta} \cdot \frac{V}{R_w}\right)^2 = m\left[\frac{\mu_0 \mu_1}{(l + \mu_1 \delta)R_w}\right]^2 nV^2 \tag{6.28}$$

6.4.3.3 Calculation of the equivalent braking radius

According to the calculating equation, the output micro-torque of the micro-motor can be expressed as
$$M = FR_{break} \tag{6.29}$$

where R_{break} is the equivalent braking radius.

Combining equations (6.23) and (6.29), M can be expressed as
$$M = FR = BI_{rot}(b - a)R_{break} \tag{6.30}$$

Because M in equations (6.25) and (6.30) are the same parameter, we get
$$BI_{rot}(b - a)R_{break} = \frac{1}{2}(b^2 - a^2)BI_{rot} \Rightarrow R_{break} = \frac{b + a}{2} \tag{6.31}$$

From equation (6.31) we conclude that the action spot of the electromagnetic force is the middle point of the rectangle, so the centroid of the rectangle is selected as the action spot. The distance between the centroid and the origin of coordinates is just the equivalent braking radius. Thus during the course of the experiment, it can conveniently be measured by vernier caliper.

6.4.4 Corresponding equipment to realize the noncontact method

6.4.4.1 Working principle of the device

A noncontact micro-torque measuring equipment is designed and fabricated using the noncontact method, simple working principle of the device is shown in Figure 6.35. The device is mainly composed of ferromagnet malleable cast-iron core with windings around it, an electronic balance (unit type: AX200, Japan; measuring range: 0~200 g; measuring precision:0.1 mg), a rounded aluminium disk, an optoelectronic transducer (0.5% relative measuring error), control circuit, AD data acquisition card and the computer. The aluminium disk is fixed on the end point of the output axis of the micro-motor. Border of the aluminium disk is placed in the gap of the ferromagnet core. On the edge of the aluminium disk, a small circle hole whose radius is about 1 mm is drilled over against the optoelectronic transducer. The bracket which is used to fix the ferromagnet malleable cast-iron core and the windings, is placed on the electronic balance. The electronic balance is connected with the computer through the serial port of the computer. In order to decrease the effect of air flow and vibration of the outer circumstance, the whole equipment is placed in an air locked protecting shield. During the course of the design turns of the windings, permitted braking potential, measuring scope of the micro-torque etc. are all preestimated by the equation in the

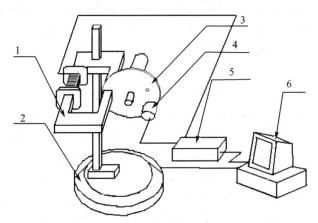

1—Ferromagnet malleable cast-iron core; 2—Electronic balance;
3—Aluminium disk; 4—Optoelectronic transducer; 5—Control circuit; 6—Computer

Figure 6.35 Working principle of noncontact micro-torque measuring equipment

theoretical calculation. When a certain DC driving is applied to the micro-motor, the micro-motor's output axis begins to rotate because of the function of the electromagnetic force between the stator and the rotor (simple testing principle of the rotating speed is shown in Figure 6.36). When the small hole is rotating across the optoelectric transducer, there is a shift of the light intensity in the optoelectric transducer. Then the varying signal of light intensity can be changed into the electronic signal and then enlarged by the control circuit. The control circuit transmitted the magnified signal to AD data acquisition card and then to the computer.

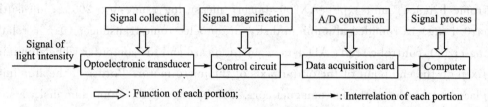

Figure 6.36 Testing principle of rotating speed

The computer changes the inputted electronic signal to the value of rotating speed by the control program and saves it in the computer. Then, a DC braking voltage is applied to the braking windings and the magnetic field is formed in the air gap. Because the aluminium disk is rotating perpendicularly to the magnetic line of force, the electromagnetic force comes into being between the aluminium and the ferromagnet core. That is just the braking force applied to the micro-motor. According to Newton's third law, a counterforce whose direction is opposite to that of the braking force is applied to the ferromagnet core and transferred to the electronic balance through the bracket. Electronic balance records the value of the braking force and transmits it to the computer through the serial port. The product of braking force and the equivalent braking radius (that is the distance between the rotating axis and the centroid of the rectangle braking magnetic pole over the aluminium disk, and can be measured by a vernier caliper) is the braking micro-torque. Then the relationship between the rotating speed and the corresponding output torque can be recorded and depicted. There are many advantages for this equipment to measure the micro-motor's micro-torque. Firstly, measuring the micro-motor rotating speed by a photoelectric sensor ensures that the high precision and easy solution of the inputted electric signal. Secondly, the real braking force is gained by the electronic balance, but not the theoretical calculated result, which possesses much larger error because of the assumptions in the calculating model. Meanwhile, serial port connecting type between the balance and the computer confirms the data to be timely transported to the computer. The transport timegap is controlled by the program easily. Thirdly, placing the system in a protecting shield decreases the effect of the outer factors. The total measuring precision is increased correspondingly.

6.4.4.2 Precision analysis of the equipment

Precision analysis of the noncontact micro-torque measuring equipment is performed using the theory of errors. There the measuring error of the device is divided into two parts, of which one is the rotating speed error and the other is the error of the output torque.

1. Error analysis of the rotating speed

In the analysis model, the total error of the rotating speed is composed of many factors such as errors of control circuit, data acquisition card and the control program which is written with VC++ etc. In order to avoid the complicated analysis, the analytical model (shown in Figure 6.37) classifies all the errors into two portions. One portion is the optoelectronic transducer's relative error, which is labeled by dashed line. The calibrated result is 0.5%. Another portion of the relative error is the processing system, which is labeled by dot-and-dash line. This portion includes the error of the control circuit, data acquisition card and the computer. Because it is very complex to analysis, the measuring precision is calibrated with a standard waveform generator. Simple calibrating theory is shown in Figure 6.38.

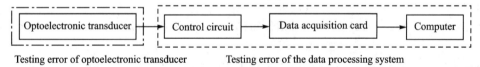

Figure 6.37 Precision analysis of the rotating speed

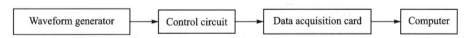

Figure 6.38 Rotating speed precision testing of the processing system

In this experiment, a high precision waveform generator (HP33120A, the output frequency of the square wave is in the range of 100 μHz ~ 100 MHz, and the frequency resolution is 100 μHz. This device can be regarded as the standard frequency output device) is used to calibrate the measuring precision of the system. During the course of the experiment, the waveform generator firstly outputs a standard square wave to simulate the signal of the optoelectronic transducer initiated by the rotating speed. At the same time, the value of the frequency is recorded by the computer. By comparing the standard output frequency and the tested result, relative error of the processing system can be obtained then. The tested results and the real standard values are shown in Table 6.2.

Table 6.2 Results of the frequency

Standard output frequency/Hz	Computer tested result/Hz	Relative error/%
10.02	10.01	0.099
99.70	99.59	0.11
199.60	199.79	0.09
399.39	399.96	0.14
597.84	597.24	0.10
798.71	798.11	0.07
1 000.15	1 001.49	0.13
1 997.05	2 000.03	0.15
4 031.01	4 027.61	0.10

From the tested results, we conclude the average value of the relative error can be expressed as

$$\bar{\delta} = \frac{\sum_{i=1}^{n} \delta_i}{n} = \frac{\sum_{i=1}^{9} \delta_i}{9} = 0.11\% \quad (6.32)$$

where $\bar{\delta}$ is the average value of the relative error, and δ_i is the relative error of a single tested result.

According to the calculating equation of limit error, the limit error of arithmetic mean for relative error can be expressed as

$$\sigma_{\bar{\delta}} = \sqrt{\frac{\sum_{i=1}^{n}(\delta_i - \bar{\delta})^2}{n(n-1)}} = \sqrt{\frac{\sum_{i=1}^{11}(\delta_i - 0.15)^2}{9 \times 8}} = 0.025\% \quad (6.33)$$

where $\sigma_{\bar{\delta}}$ is the limit error of arithmetic mean.

The real relative error of the rotating speed can be expressed as

$$\delta = \bar{\delta} \pm 3\sigma_{\bar{\delta}} = 0.11\% \pm 0.07\% \quad (6.34)$$

The precision of the rotating speed which is the synthetical relative error of the optoelectronic transducer and the processing system can be expressed as

$$\delta_{\text{rot}} = \sqrt{\delta^2 + \delta_{\text{sensor}}^2} = \sqrt{0.11^2 + 0.5^2} = 0.51\% \quad (6.35)$$

where δ_{rot} is the total relative error of the speed testing system, δ_{sensor} is the relative measuring error of the optoelectronic transducer. This calculated value is precise enough to carry out the speed testing.

2. Precision analysis of the tested micro-torque

According to equation (6.29) in the theoretical calculation, output micro-torque is the product of the balance inputted result and the braking radius, and it can be expressed as

$$T = FR \qquad (6.36)$$

where T represents the tested micro-torque of the device, F is the braking force that is applied on the electronic balance, R is the braking radius. According to the theory of errors, the micro-torque error can be expressed as

$$\delta T = \sqrt{\left(\frac{\partial T}{\partial F}\right)^2 \delta F^2 + \left(\frac{\partial T}{\partial R}\right)^2 \delta R^2} \qquad (6.37)$$

where δF is the measuring error of the electronic balance, and δR is the measuring precision of the aluminium disk's braking radius. Their real values are 0.1 mg and 0.01 mm. Because $\frac{\partial T}{\partial F}=R$ and $\frac{\partial T}{\partial R}=F$, equation (6.37) can be substituted as

$$\delta T = \sqrt{(R \cdot \delta F)^2 + (F \cdot \delta R)^2} \qquad (6.38)$$

From equation (6.38) we conclude that if R and F are measured in a certain experiment, the corresponding measured micro-torque error can be calculated.

6.4.5 Experiment result and discussion

A planar DC brushless micro-motor, which is fabricated by MEMS bulk technology and quasi-LIGA process (the fabricated figure of the stator is shown in Figure 6.39), is tested with the non-contact micro-torque testing equipment. For this experiment, the micro-motor's rotating speed is in the range of 5 000~7 000 rpm, which is most commonly used for the micro-torque. The braking radius measured by the vernier caliper is 12 mm and the measuring precision is 0.01 mm. During the course of the experiment the micro-motor is firstly rotated without any load applied on it. Then the gradually increased DC potentials are applied on the windings. Every rotating speed and the micro-torque corresponding to the driving voltage applied on the windings are listed in Table 6.3.

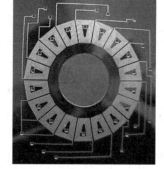

Figure 6.39 Fabricated sketch of the stator

According to equation (6.36), the calculated micro-torques corresponding to the rotating speeds are also listed in the third column of Table 6.3. Measuring precision calculated with equation (6.38) of a single micro-torque is listed in the fourth column. Thus the average testing precision of output torque can be expressed as

$$\overline{T_\delta} = \frac{\sum_{i=1}^{n} T_{\delta i}}{n} = \frac{\sum_{i=1}^{11} T_{\delta i}}{11} = 0.01 \ \mu N \cdot m \qquad (6.39)$$

Table 6.3 Tested results of the micro-torque

Rotating speed/rpm	Tested result of the electronic balance/mg	Calculated output torque/($\mu N \cdot m$)	Calculated measuring precision/($\mu N \cdot m$)
5 000	1 799.3	216.09	0.012 20
5 200	1 742.7	209.29	0.012 19
5 400	1 672.8	200.9	0.012 18
5 600	1 595.3	191.59	0.012 16
5 800	1 524.6	183.12	0.012 15
6 000	1 457.9	175.09	0.012 14
6 200	1 406.3	168.89	0.012 13
6 400	1 335.5	160.39	0.012 12
6 600	1 271.4	152.69	0.012 11
6 800	1 205.7	144.81	0.012 10
7 000	1 134.9	136.31	0.012 09

According to the theory of errors, the standard deviation of arithmetic means can be expressed as

$$\sigma_{\overline{T_\delta}} = \sqrt{\frac{\sum_{i=1}^{n}(T_{\delta_i} - \overline{T_\delta})^2}{n(n-1)}} = \sqrt{\frac{\sum_{i=1}^{11}(\delta_i - 0.01)^2}{11 \times 10}} = 4 \times 10^{-5} \ \mu N \cdot m \qquad (6.40)$$

According to the 3σ principle, the final calculated measuring precision can be expressed as

$$T_\delta = \overline{T_\delta} \pm 3\sigma_{\overline{T_\delta}} = (1 \pm 0.01) \times 10^{-2} \ \mu N \cdot m \qquad (6.41)$$

From the calculated result, we can conclude that the finally calculated measuring precision is $0.01 \ \mu N \cdot m$, which is precise enough for the output torque test of the micro-motor.

6.5 Microscopic morphology testing method

One of the typical characteristics of MEMS is its functional size in microns, and its measured force magnitude is generally in mN even nN, so it is infeasible to test its morphology and mechanical properties with conventional test methods. Especially when doing the testing of the surface topography to assess its machining effect, the conventional testing equipment seems unsuitable. Therefore, we need a specialized testing equipment to test its performance. This section describes the most commonly used equipment for the testing of micro-force and surface morphology—AFM (Atomic Force Microscope).

The working principle and data processing method are introduced as follows.

AFM is the most commonly used equipment for the testing of micro-force and surface morphology at home and abroad. In order to extract the tribology data, the equipment mainly depends on the laser beam deflection method. Figure 6.40 shows the principle of AFM tribology test. The testing process describes as placing the samples in AFM to form a contact pair with the pinpoint. The AFM pinpoint is mounted on a V-shaped cantilever which is very sensitive to the weak force, and the other end of the cantilever is fixed. By adjusting the position of the sample stage, the pinpoint approaches the

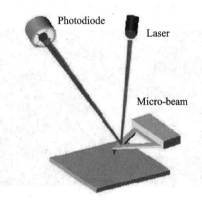

Figure 6.40 Principle of AFM tribology test

sample surface (the distance is close to the atomic radius). Thus, van der Waals force will initiate between the sample surface and the pinpoint. Then, gently slide the pinpoint on the sample surface, and during the sliding, the force between the top atoms and the atoms on the sample surface will bend the cantilever from the original position. The offset can be measured by a pair of photodiodes. Then, convert the cantilever bending deformation signals into optical signals and amplify the signals, and we can obtain the surface topography data. Compared with other forms of surface topography test, this method can achieve nano-sized morphology testing. At the same time, by calculating the product of the relative displacement and the stiffness of cantilever pinpoint, we can get the value of the measured micro-force. For the testing device, the core sensitive element is the gilded pinpoint produced by a MEMS etching and electroplating process.

In the MEMS system, different surface topography and measurement conditions (e. g. testing temperature, air humidity etc.) will lead to completely different results. Therefore, in the testing process, it's necessary to record the specific temperature, humidity etc., and ensure the complete isolation with the external noise. During the test, the scan area of the pinpoint is 5 μm$\times$5 μm, the pinpoint needs to acquire 256 data along each x or y direction, and totally 65 536 data. Then by a specific data processing technology we can get the reduction of the relevant surface topography and obtain all the data. Figure 6.41 and Figure 6.42 show two typical MEMS device surface topography testing results (the top surface and the side surface of the component). By using the software pre-installed in AFM, we can get the coordinates of each test point and corresponding height. By using MATLAB to calculate the height of each collection point, we can get its topography. The detailed computing method of each index shows in Subsection 5.3.4.

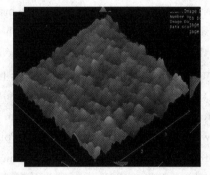

Figure 6.41　Sketch of the lateral surface　　　**Figure 6.42**　Sketch of the top surface

6.6　Summary

This chapter takes the MEMS testing technology as the study object, and gives a detailed introduction on the acceleration, angular velocity, pressure, micro-moment, surface morphology and tiny forces testing principle, some corresponding theories and the development status, also introduces the mathematical models.

References

[1] He Tiechun. Inertial Navigation Accelerometer [M]. Beijing: National Defense Industry Press, 1983.

[2] Li Baozhang, Yuan Gannan. Some Problems on The Static Balance of The Floating Pendulum for The Accelerometer [J]. Chuan Gong Ke Ji,1977(4):13-21.

[3] Wang Guangda. Analysis and Design of A Vibration Wire Accelerometer [J]. Automation Instrumentation, 1985(1):7-10.

[4] Sun Yusheng. Parameter Calculation and Error Analysis for The Quartz Flexure Accelerometer[J]. Journal of Vibration, Measurement and Diagnosis, 1984(3):13-22.

[5] Ding Henggao. Micro-nano Technology-army-civilian Application Technology for 21st century[J]. Chinese Journal of Scientific Instrument, 1995,16(1):1-7.

[6] Zimmermann L, Ebrsohl J Ph. Airbag Application: A Microsystem Including A Silicon Capacitive Accelerometer, COMS Switched Capacitor Electronics and True Self-test Capability [J]. Sensors and Actuators A, 1995, 46:190-195.

[7] Harvey Weinberg. Application of MEMS Motion Sensor on The Moving Mobile[J]. Electric World,2006(10):75-77.

[8] Qian Pengan. Application of Accelerometer on The Measurement of Human Movement [C]//Application and Development of Computer Technology Conference, Hefei, Anhui. 2004:632-636.

[9] Zheng Li. Fall Detection and Reliable Device Design [J]. Chinese Journal of Medical Instrumentation,2009,33(2):99-102.

[10] Willis J, Jimerson B D. A Piezoelectric Accelerometer [J]. Proc. IEEE, 1964, 52(7): 871-872.

[11] Satchell D W, Greenwood J C. A Thermally-excited Silicon Accelerometer [J]. Sensors and Actuators, 1989(17):241-245.

[12] Chang C S, Putty M W. Resonant-bridge Two-axis Microaccelerometer [J]. Sensors and Actuators A,1990(21-23):342-345.

[13] Cheshmehdoost A, Jones B E, Oconnor B. Characteristics of A Force Transducer Incorporating A Mechanical DETF Resonator [J]. Sensors and Actuators A, 1991(25-27):307-312.

[14] Cheshmehdoost A, Stroumbouhs S. Dynamic Characteristics of A Resonating Force Transducer [J]. Sensors and Actuators A, 1994(41-42):74-77.

[15] Cheshmehdoost A, Jones B E. Design and Performance Characteristics of An Integrated High-capacity DETF-based Force Sensor [J]. Sensors and Actuators A, 1996(52):99-102.

[16] Honeywell. RBA500 Datasheet. https://aero1.honeywell.com/inertsensor/docs/rba500.pdf.

[17] Barbour N, Schmidt G. Inertial Sensor Technology Trends [J]. IEEE Sensors Journal, 2001, 4(1):146-151.

[18] Ohwada K, Mochida Y, Kawai H. Micromachined Silicon Gyroscope [J]. Sensors Update, 1999, 6(1):95-116.

[19] Stephane D, Dick D, Tony M, et al. MEMS Gyro for Space Applications Overview of European Activities. AIAA Guidance, Navigation, and Control Conference. San Francisco, USA, 2005: 6232-6243.

[20] Najafil K, Junseok C, Haluk K, et al. Micromachined Silicon Accelerometers and Gyroscopes. IEEE/RSJ International Conference on Intelligent Robots and Systems. Las Vegas, USA, 2003: 2353-2358.

[21] Greiff P, Boxenhorn B, King T, et al. Silicon Monolithic Micromechanical Gyroscope [A]. Technical Digest of The 6th International Conference on Solid-State Sensors and Actuators (Transducers '91) [C]. San Francisco, USA, 1991:966-968.

[22] Bernstein J, Cho S, King A T, et al. A Micromachined Comb-Drive Tuning Fork Rate Gyroscope. Proceedings of IEEE Micro Electro Mechanical Systems Workshop (MEMS'93). Fort Lauderdale, FL, 1993: 143-148.

[23] Weinberg M, Bernstein J, Cho S, et al. A Micromachined Comb-Drive Tuning Fork Gyroscope for Commercial Applications. Proceedings of Sensor Expo. 1994:187-193.

[24] Hopkin I D. Vibrating Gyroscopes (Automotive Sensors). IEEE Colloquium on Automotive Sensors. Digest No. 1994/170, Solihull, UK, 1994: 1-4.

[25] Tanaka K, Mochida Y, Sugimoto M, et al. A Micromachined Vibrating Gyroscope [J]. Sensors and Actuators A: Physical, 1995, V50(2): 111-115.

[26] Tanaka K, Mochida Y, Sugimoto S, et al. A Micromachined Vibrating Gyroscope. Proceedings of Eighth IEEE Int. Conf. on Micro Electro Mechanical System (MEMS'95). Amsterdam, Netherlands, 1995: 278-281.

[27] Clark W A, Howe R T, Horowitz R. Surface Micromachined Z-axis Vibratory rate Gyroscope. Technical Digest, Solid-State Sensor and Actuator Workshop. Hilton Head Island, SC, USA, 1996: 283-287.

[28] Juneau T, Pisano A P, Smith J H. Dual Axis Operation of a Micromachined Rate Gyroscope. Proceedings of International Conference on Solid-State Sensors and Actuators. Transducers '97, Chicago, USA, 1997: 883-886.

[29] Oh Y, Lee B, Baek S, et al. A Surface Micromachined Tunable Vibratory Gyroscope. Proceedings of IEEE Micro Electro Mechanical System Workshop (MEMS'97). Japan, 1997: 272-277.

[30] Lutz M, Golderer W, Gerstenmeier J, et al. A Precision Yaw Rate Sensor in Silicon Micromachining. Proceedings of The 9th International Conference on Solid-State Sensors and Actuators. Transducers 1997, Chicago, USA, 1997: 847-850.

[31] Mochida Y, Tamura M, Ohwada K. A Micromachined Vibrating Rate Gyroscope with Independent Beams for The Drive and Detection Modes. Proceedings of The 12th IEEE Int. Conf. on Micro Electro Mechanical Systems (MEMS'99). Orlando, FL, 1999: 618-623.

[32] Mochida Y, Tamura M, Ohwada K. A Micromachined Vibrating Rate Gyroscope with Independent Beams for The Drive and Detection Modes [J]. Sensors and Actuators A: Physical, 2000,80(2):170-178.

[33] Xie H, Fedder G K. A CMOS-MEMS Lateral-axis Gyroscope. Proceedings of The 14th IEEE Int. Conf. Microelectromechanical Systems. Interlaken, Switzerland, 2001: 162-165.

[34] Luo H, Zhu X, Lakdawala H, et al. A Copper COMS-MEMS Z-axis Gyroscope. Proceedings of The 15th IEEE Int. Conf. Micro Electro Mechanical Systems. Las Vegas, NV, USA, 2001:631-634.

[35] Geiger W, Butt W U, Gaisser A, et al. Decoupled Microgyros and the Design Principle DAVED [J]. Sensors and Actuators A: Physical, 2002, 95(3):239-249.

[36] Choi B-d, Park S, Ko H, et al. The First Sub-deg/hr Bias Stability, Silicon-microfabricated Gyroscope. Proceedings of Digest of Technical Papers Solid-State Sensors, Actuators and Microsystems TRANSDUCERS '05, The 13th International Conference. 2005,1: 180-183.

[37] Alper S E, Akin T. A Symmetric Surface Micromachined Gyroscope with Decoupled Oscillation Modes [J]. Sensors and Actuators A: Physical, 2002,97(8):347-358.

[38] Lin P Qi, Stern. Analysis of a Correlation Filter for Thermal Noise Reduction in a MEMS Gyroscope. Proceedings of The 34th Southeastern Symposium on System Theory. 2002: 197-203.

[39] Wang C-h, Huang X. Application of Wavelet Packet Analysis in the De-noising of MEMS Vibrating Gyro. Proceedings of 2004 Position Location and Navigation Symposium. 2004: 129-132.

[40] Dong L, Leland R. The Adaptive Control System of A MEMS Gyroscope with Time-varying Rotation Rate. Proceedings of 2005 American Control Conference. 2005,5: 592-3597.

[41] Singwi K S, Udgaonkar B M. Piezoresistance Effect in Germanium and Silicon [J].

Physical Review, 1954, 94(1): 42-49.

[42] Greenwood J C. Etched Silicon Resonant Sensor[J]. Journal of Physics E: Scientific Instruments, 1984, 17: 650-652.

[43] Greenwood J C. Miniature Silicon Resonant Pressure Sensor [J]. IEEE Proceedings D on Control Theory and Applications, 1988, 135(5): 369-372.

[44] Petersen K, et al. Resonant Beam Pressure Sensor Fabricated with Silicon Fusion Bonding. International Conference on Solid-State Sensors and Actuators, San Francisco, CA, USA, 1991(b): 664-667.

[45] Parsons P, Glendinning A, Angelidis D. Resonant Sensors for High Accuracy Pressure Measurement Using Silicon Technology [J]. IEEE AES Magazine, 1992(7): 45-48.

[46] Ikeda K, et al. There-dimensional Micromachining of Silicon Pressure Sensor Integrates Resonant Strain Gauge on Diaphragm [J]. Sensors and Actuators A, 1990, 21-23: 1007-1010.

[47] Saigusa T, Kuwayama H. Intelligent Differential Pressure Transmitter Using Microresonators. International Conference on Industrial Electronics, Control, Instrumentation and Automation, 1992,(9-13)3: 1634-1639.

[48] Ehrfeld Nienhaus W, Schmitz F, et al. Design and Realization of a Penny-shaped Micromotor. Proceedings of the SPIE, 1999, 36(80):592-600.

[49] Gilles P A, Delamare J, Cugat O, et al. Design of A Permanent Magnet Synchronous Micromotor. The 35th ISA Annual Meeting and World Conference on Industrial Application of Electrical Energy. 2000: 223-227.

[50] Livermore Carol, Forte Anthony R, Lyszczarz Theodore, et al. A High-Power MEMS Electric Induction Motor[J]. Journal of Microelectromechanical System, 2004, 13(3): 465-471.

[51] Mathiesont D, Robertsont B J, Beerschwingert U, et al. Micro Torque Measurement for A Prototype Turbine[J]. Journal of Micromechanics and Microengineering, 1994 (4),4:129-139.

[52] Gass V, Schoot B H van der, Jeanneret S, et al. Micro-torque Sensor Based on Differential Force Measurement. Proceedings of IEEE Micro Electro Mechanical Systems,1994:241-244.

[53] Brenner W, Haddad G, Detter H, et al. The Measurement of Mini-motors and Micromotors Torque-characteristic Using Miniaturized Cable Brake[J]. Microsystem Technology, 1997:68-71.

[54] Son D, Lim S J, Kim C S. Non-contact Torque Sensor Using The Difference of

Maximum Induction of Amorphous Cores. IEEE Transactions on Magnetics, 1992, 28 (5): 2205-2207.

[55] Ota1 H, Ohara T, Karata Y, et al. Novel Micro Torque Measurement Method for Microdevices [J]. Journal of Microelectromechanical System, 2001 (11): 595-602.

[56] Ota1 Hitoshi, Li Luming, Takeda Munehisa, et al. Torque Measurement Method Using Air Turbine for Micro Devices. Proceedings of IEEE Micro Electro Mechanical System (MEMS), 2000: 223-228.

Chapter 7
Application examples of the finite element method in the design of MEMS devices

The MEMS devices have long processing cycles, high processing costs and problems of batch processing (the unreasonable design of certain parts of the component will lead to the processing failure for all devices) in the design and manufacturing process. Therefore, it is very important to take a pre-assessment for the structural performance of design. Using the finite element technique, simulating the parameters of the agencies designed, finding the deficiencies in the design before processing and putting forward reasonable proposals, is one of the most effective methods.

Currently, the finite element software commonly used internationally includes Ansys, Ideas, Conventorware and IntelliSuite etc. Among them, Ansys is the commercial software early used for finite element analysis of MEMS. It can be used for the analysis of mechanics, thermal, electromagnetic of the structures, the analysis of fluid and multi-physics coupling etc. It is the main simulation tool of many research units.

This chapter takes this software as the study object and makes detailed explanation of its application in MEMS simulation with a large number of engineering practices.

7.1 Important concepts of the software

1. Element types

Element types are to tell the computer what kind of unit model will be used for the analysis of problem. The element types include the particles and voids units of 0-D, beams and columns units of 1-D, plates and shells units of 2-D, tetrahedral and hexahedral units of 3-D, and so on.

The method of adding element types is Preprocessor→Element Type→Add/Edit/Delete.

Enter the interface shown in Figure 7.1, then you can add element types in it.

2. Real constants

Real constants are to define the section characteristics of model under analysis. In the beam or truss model, you must define its cross-sectional dimension. In the film or plate model, the thickness must be defined. If you do not enter the section properties, the analysis can't be carried out. The size of three-dimensional structure includes all parameter information, so do not need to define the relevant parameters in the Real Constants.

3. Material properties

Before finite element analysis, you must define some properties of the material. The parameters which are generally needed to be defined of the material properties section include the elastic modulus EX and EY, the shear modulus GXY, the Poisson's ratio NUXY, the thermal expansion coefficient and density and so on. For thermal analysis, electromagnetic analysis and piezoelectric analysis, some parameters such as thermal conductivity, resistivity and dielectric constant are needed be entered. In general, when the material is a constant, the parameter is a constant. In shaping analysis, the material property curves are also needed to be entered such as the change curve of elastic modulus with temperature. In gradient analysis, you will need to enter the time curve of the material. The material density is needed to be entered in vibration analysis.

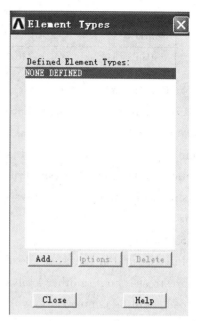

Figure 7.1 Adding interface of element types

In MEMS analysis, because it is the dimensionless input, the calculation results are determined by the material parameters which are inputted. When the unit of the size inputted is meter, the unit of other parameters such as Young's modulus should be Pa, and the unit of the density should be kg/m^3. When other parameters follow the international units, the results calculated will follow standard international units too. For example, the unit of displacement results is meter, the force unit is Newton and the stress unit is Pa and so on. When the value is analyzed in microns, the corresponding material properties are shown in Table 7.1.

Table 7.1 Material properties

Item	Size	Elastic Modulus	Density	Displacement	Force	Frequency
Unit	μm	MPa	$kg/\mu m^3$	μm	μN	Hz

4. Constraints

Constraint is the method that defines the fixed part of a structure. Before analyzing a structure, its boundary conditions must be properly defined. In the structural analysis, the boundary conditions are defined as the displacement and torque along the x, y and z directions. Different analysis types have different constraints. For example, both the temperature in temperature analysis and the voltage in electromagnetic analysis could be used as the constraints.

5. Loads

In structural analysis, the loads include fixation forces, distributed loads, accelerations and prestresses etc. In other analyses such as thermal analysis and electromagnetic analysis, the load could be temperature, voltage and so on.

7.2 Introduction of the Ansys software interface

The operation interface of Ansys is shown in Figure 7.2. It mainly includes Main Menu, Utility Menu, Input Window, Toolbar and Output Window etc. The Ansys interface is shown in Figure 7.2.

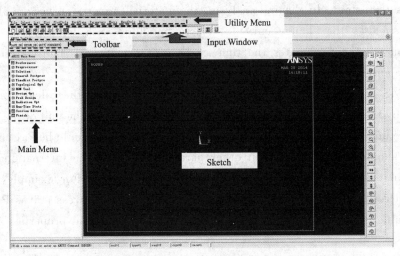

Figure 7.2 Ansys interface

The functions of each window are described below.

Utility Menu: It is the main application menu of Ansys. Its main functions are file operations, selection, graphical control and the establishment of parameters etc. It includes 10 pull-down menus: File, Select, List, Plot, PlotCtrls, WorkPlane, Parameters, Macro, MenuCtrls and Help. Its operation is similar to the Windows system menu.

Input Window: In Ansys operation, there is not only the GUI mode but also the command stream mode. This window is used to enter the command line. After input a command or data, only when you press the Enter key on the keyboard, the command or data will be received by the system. In addition, this window can also display the message inputted in GUI and browse commands entered previously.

Toolbar: It consists of a set of buttons which perform the common commands of Ansys. The common commands could be set as the Toolbar button for easy call. Several default buttons are as follows: SAVE_DB, RESUM_DB, QUIT, POWRGRPH, E-CAE.

Graphic Output: Display the user-created models in Ansys or graphics imported by other software, and complete all graphics pick-up operation. In addition, this window can also display grid, calculation results, contours, isoline and other graphics.

Output Window: Display all text output associated with Ansys program commands, warning and error information etc. The Output Window is usually located behind other windows. If you want to see the output of Ansys, you could make the Output Window show in front of other windows by clicking it. Especially during non-linear calculations, we often view the calculation process and results through the Output Window.

Main Menu: This window contains the main features of Ansys such as pre-treatment, solution and post-processing. It contains a total of 12 menu items: Preferences, Preprocessor, Solution, General Postprocessor, TimeHist Postprocessor, Topological, Design OPT, Probe Designing, Radiation OPT, Run-Time Stats, Session Editor and Finish.

7.3 The coordinate system in Ansys

Before the modeling, it is necessary to understand the concepts of coordinate system and working plane in Ansys. The skillful use of various coordinate systems will greatly facilitate the users. Several typical coordinate systems in Ansys are described below.

Global coordinate system: Define the spatial position of the geometry parameters such as node and keypoints.

Local coordinate system: Custom coordinate system that has the same role as global coordinate system.

Node coordinates: Define the direction of degrees of freedom and result data of each node.

Display coordinate system: Used for the list and display of geometry parameters.

Element coordinate system: Determine the direction of the spindle of material properties and the cell result data.

Results coordinate system: In the general post-processing operation, it could be used to convert the node or unit results from one coordinate system to another specific coordinate system.

Although there are a variety of coordinate systems in Ansys and the user can define multiple coordinate systems, only one coordinate system could be activated at the same time. System always activates the Cartesian coordinate system first. If the user defines a new local coordinate system, the new coordinate system will be automatically activated.

Ansys could activate any coordinate system at any time while the program is running. If not explicitly change the activated coordinate system, the current active coordinate system will remain in effect.

The following methods can activate the coordinate system.

GUI: Utility Menu | WorkPlane | Change Active CS to | Global Cartesian
Utility Menu | WorkPlane | Change Active CS to | Global Cylindrical
Utility Menu | WorkPlane | Change Active CS to | Global Cylindrical y
Utility Menu | WorkPlane | Change Active CS to | Global Spherical
Utility Menu | WorkPlane | Change Active CS to | Specified Coord Sys⋯
Utility Menu | WorkPlane | Change Active CS to | Working Plane

When define the nodes or keypoints, the program will display the coordinates as x, y and z regardless of which coordinate system is activated currently. If the current is not a Cartesian coordinate system, the user should understand the x, y and z as the R, θ and Z in cylindrical coordinates or the R, θ and Φ in spherical coordinates.

Each coordinate system is described below.

1. The global coordinate system

The global coordinate system is used to determine the position of the space geometrical structures. It is an absolute reference system, providing a reference point for all coordinates. There are three kinds of coordinate systems for you to choose (Cartesian coordinates, cylindrical coordinates and spherical coordinates). They are all right-handed coordinate system and have a common origin. In the definition of a critical point or node, it is usually

Chapter 7 Application examples of the finite element method in the design of MEMS devices

more convenient when use the Cartesian coordinate. Meanwhile, it is more convenient to use the cylindrical coordinates or spherical coordinates when define the structure with an arc. In Ansys, these three coordinates are identified by numbers: 0 represents Cartesian coordinates, 1 represents cylindrical coordinates and 2 for spherical coordinates. No matter what coordinate system is used now, if you want to go back to the global coordinate system, the operation is as follow:

Utility Menu | WorkPlane | Change Active CS to | Global Cartesian
Utility Menu | WorkPlane | Change Active CS to | Global Cylindrical
Utility Menu | WorkPlane | Change Active CS to | Global Cylindrical y
Utility Menu | WorkPlane | Change Active CS to | Global Spherical

2. The local coordinate system

In Ansys, the local coordinate system is mainly used for graphics copy, paste, mirror and other operations. In general, the processes of graphic copy, paste and mirror take the generic coordinates as their default work coordinate system. Therefore, in the course of these operations, the results may not meet the design requirements, so the required local coordinate system is needed to be established in operations. Define the current work coordinate system as the local coordinate system and the copy and paste operations can be realized in accordance with the expected.

Overall Cartesian coordinate defines the local coordinate system.

  GUI: Utility Menu | WorkPlane | Local Coordinate Systems | Create Local CS |
    At Specified Loc

Define the local coordinate system by an existing node.

  GUI: Utility Menu | WorkPlane | Local Coordinate Systems | Create Local CS |
    By 3 Nodes

Define the local coordinate system by the existing keypoints.

  GUI: Utility Menu | WorkPlane | Local Coordinate Systems | Create Local CS |
    By 3 Keypoints

Define the local coordinate system by taking the origin of the current work plane as the center.

  GUI: Utility Menu | WorkPlane | Local Coordinate Systems | Create Local CS |
    At WP Origin

3. The work plane

The work plane is a reference plane which can be moved, and it is equivalent to a virtual drawing board. All entity models must be generated on the work plane. Although the cursor is only showed as a point on the screen, it actually represents a line perpendicular to the surface in space. In order to use the cursor to pick up a point, you must use an imaginary plane to intersect the straight line. Only in this way could you determine the unique point in space. This imaginary plane is the work plane. Imagine the relationship between cursor and work plane from another point of view, the cursor could be described as a point wandering in the work plane. Therefore, the work plane is just as a tablet which you could write on. The work plane may not be parallel to the plane of the display. The work plane is an infinite plane with the origin, the two-dimensional coordinate system, capture incremental and display grid. Only one work plane can be defined at the same time. When you define a new work plane, the existing work plane will automatically be deleted. The work plane is independent to coordinate system. For example, the work plane may have different origin and direction of rotation with the activation coordinate system. After entering the Ansys, the system has a default work plane, the x-y plane of the overall Cartesian coordinate system (rectangular coordinate system). The x-axis and y-axis of the coordinate system is respectively the wx axis and wy axis of the work plane.

(1) Control the display and style of the work plane.

To display the status of the work plane including the location, orientation, and incremental, you can use the following methods:

  GUI:Utility Menu | List | Status | WorkPlane

  Command: WPSTYL, STAT

To restore the default settings work plane, you can use the command: WPSTYL, DEFA.

(2) Define a new work plane.

Users can use one of the following five methods to define a new work plane.

Define a work plane by three points or define the plane which is over a specified point and perpendicular to the visual vector as the work plane.

  GUI:Utility Menu | WorkPlane | Align WP with | xyz Locations

  Command: WPLANE

Define a work plane by three keypoints or define the plane which is over a specified key point and perpendicular to the visual vector as the work plane.

  GUI: Utility Menu | WorkPlane | Align WP with | Keypoints

  Command: KWPLANE

Define the plane over a point on the designated line and perpendicular to the visual vector as the work plane.

GUI: Utility Menu | WorkPlane | Align WP with | PlaneNormal to Line

Command: LWPLANE

Define the x-y (or R-θ) plane in the existing coordinate system as the work plane.

GUI: Utility Menu | WorkPlane | Align WP with | Active Coord Sys

Utility Menu | WorkPlane | Align WP with | Global Cartesian

Utility Menu | WorkPlane | Align WP with | Specified Coord Sys...

Command: WPCSYS

4. Move and rotate the work plane

By default, the working plane coincides with the overall coordinate plane. But in the process of creating a 3-D geometric model, the user can move the work plane to the desired location. Meanwhile, in the creation of 3-D solid models, the work plane is often needed to be rotated to ensure that the entities are operated in the work plane. Move the origin of the work plane to the middle position of the keypoints.

GUI: Utility Menu | WorkPlane | Offset WP to | Keypoints

Command: KWPAVE

Move the origin of the work plane to the middle position of the nodes.

GUI: Utility Menu | WorkPlane | Offset WP to | Nodes

Command: NWPAVE

Move the origin of the work plane to the middle position of the designated point.

GUI: Utility Menu | WorkPlane | Offset WP to | Global Origin

Utility Menu | WorkPlane | Offset WP to | Origin of Active CS

Utility Menu | WorkPlane | Offset WP to | xyz Locations

Command: WPAVE

Shift the work plane incrementally.

GUI: Utility Menu | WorkPlane | Offset WP by Increments

Command: WPOFFS

Rotate the work plane.

GUI: Utility Menu | WorkPlane | Offset WP by Increments

Command: WPROTA

WPSTYL command or GUI method can enhance the function of the work plane. They enable the work plane to have the function of capture increment, display grid, recovery tolerance and coordinate type. In addition, it allows the user's coordinate system move with the movement of the work plane.

GUI: Utility Menu | WorkPlane | Change Active CS to | Global Cartesian

Command: CSYS

5. Restore the defined work plane

In ANSYS, the work plane cannot be stored. However, the user can create a local coordinate system at the origin of the work plane to restore the defined work plane.

Create a local coordinate system at the origin of the work plane.

    GUI: Utility Menu | WorkPlane | Local Coordinate Systems | Create Local CS |
        At WP Origin

    Command: CSWPLA

Use the local coordinate system to restore a defined work plane.

    GUI: Utility Menu | WorkPlane | Align WP with | Active Coord Sys
        Utility Menu | WorkPlane | Align WP with | Global Cartesian
        Utility Menu | WorkPlane | Align WP with | Specified Coord Sys

Command: WPCSYS

7.4 Engineering examples

This section takes the silicon material which is most commonly used in MEMS as the basis, and the most commonly used elements in MEMS including single-clamped beams, double-clamped beams, comb capacitance drive and diaphragm as the analysis objects, to take the application analysis of the actual project, such as static analysis, modal analysis, harmonic analysis and fatigue analysis, and give a detailed explanation to the relevant skills of calculation.

7.4.1 Static analysis of single-clamped beam

7.4.1.1 Static analysis of silicon material

Single-clamped beam is widely used in MEMS, including the AFM and various force sensors etc. This section takes the single-clamped beam as an example, to explain its related operations which are important.

1. Operating examples

As shown in Figure 7.3, there is a MEMS single-clamped beam (fixed at left) which is based on single crystal silicon material. Its length is 100 μm, width is 10 μm and thickness is 4 μm. When apply a vertical downward concentration force of 20 μN in the middle (as shown in Figure 7.3) of its end, we can calculate:

(1) The displacement of the rightmost of beam;
(2) The distribution of stress at the midline of the beam surface along the beam.

Mechanical parameters of monocrystalline: the elastic modulus is 169 GPa; the Poisson's ratio is 0.25; the density is 2 330 kg/m^3.

Description: since the size is in micrometer, all the material properties in the modelling process should be modelled in accordance with microns.

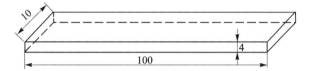

Figure 7.3 Single-clamped silicon beam structure diagram (unit: μm)

2. Calculation steps

(1) Preprocessor

● Add a element type:

Preprocessor → Element Type → Add/Edit/Delete → Add → Solid → Brick 8 Node 45 → Close, add to Solid45 unit.

● Add material properties:

Preprocessor → Material Props → Material Models, enter the interface shown in Figure 7.4.

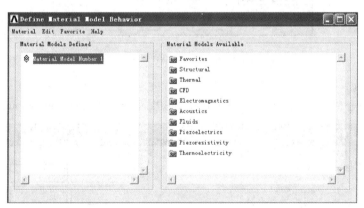

Figure 7.4 Material properties

Structural → Linear → Elastic → Isotropic, enter the interface of material properties. Fill the elastic modulus and the Poisson's ratio corresponding to the micron units (shown in Figure 7.5).

Similarly, Structural→Density, add density information (shown in Figure 7.6).

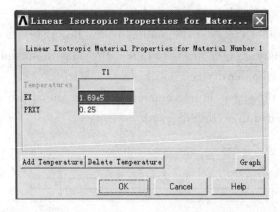

Figure 7.5 Material mechanical parameters

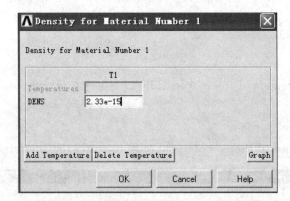

Figure 7.6 Density parameter

- Modelling:

Preprocessor→Modelling→Create→Volumes→Block→By Dimension (shown in Figure 7.7).

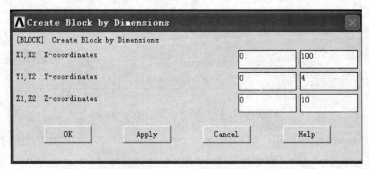

Figure 7.7 Beam modelling

The finished model is shown in Figure 7.8.

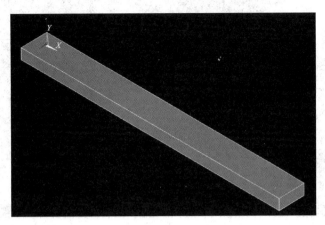

Figure 7.8 Single-clamped beam model

● Meshing:

Preprocessor → Meshing → MeshTool, enter the user interface shown in Figure 7.9. Select Global→Set →Element Edge Length, enter a value of 0.5.

Select Shape→Hex→Sweep, sweep and generate mesh, as shown in Figure 7.10.

(2) Solution

● Set the type of analysis:

Solution→Analysis Type→New Analysis→Static.

● Impose constraints:

Solution → Define Loads → Apply → Structrual → Displacement→On Areas, select the leftmost end face, then select all DOFs, and the value is 0.

● Add a concentrated load:

Solution → Define Loads → Apply → Structrual → Force/Movement→On Nodes.

Select the midpoint of the upper end face, the direction is down along the y axis, the value is 20 μN (shown in Figure 7.11).

Calculation: Solution→Solve→Current LS.

(3) General Postproc

Figure 7.9 Meshing interface

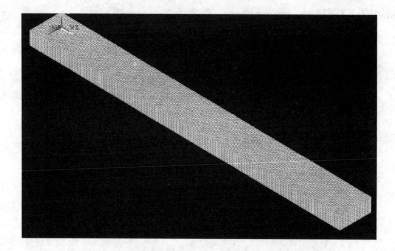

Figure 7.10 Generated mesh

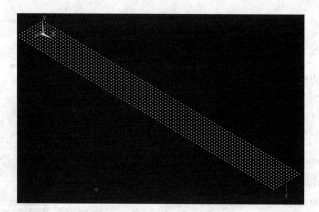

Figure 7.11 Load chart

● Check the stress distribution:

General Postproc→Plot Results→Contour Plot→Nodal Solution, enter the interface shown in Figure 7.12. Select the equivalent stress calculated item, click OK.

Get the stress distribution of beam which is shown in Figure 7.13. The figure shows that the maximum stress is 249 MPa at the action point of concentrated force.

● Deformed shape:

General Postproc→Plot Results→Deformed Shape, select Def + Undef Edge, click OK and get the displacement map shown in Figure 7.14. The maximum value is 0.738 μm.

● View the stress distribution data of beam:

Chapter 7　Application examples of the finite element method in the design of MEMS devices

Figure 7.12　Stress choice

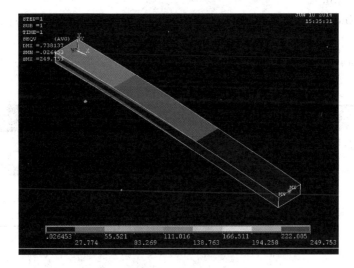

Figure 7.13　Calculated results

General Postproc→List Results→Contour Plot→Nodal Solution, select stress data.

In many cases, taking the finite element analysis is to get a better look at its parameters distribution (such as stress and displacement distribution along the beam) which is needed to

231

Figure 7.14 Displacement map of beam

be shown in curve. This request can be met by using path operations method. The steps are as follows.

Step 1: create a path.

General Postproc → Path Operation → Define Path → By Nodes, select the midpoint of both ends of the upper surface (shown in Figure 7.15).

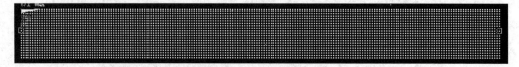

Figure 7.15 Path definition

After selection, draw the following interface. The path is named as Guo.

Step 2: call path.

You can define many paths. If you want to call one of the paths, the specific steps are as follows:

General Postproc → Path Operation → Recall Paths, select the desired path shown in Figure 7.16.

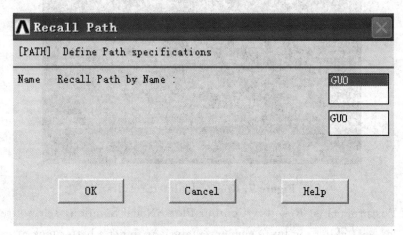

Figure 7.16 Path selection

Chapter 7 Application examples of the finite element method in the design of MEMS devices

Step 3: select the items (stress, displacement etc.) to display on the path.
General Postproc→Path Operation→Map Onto Path, get the interface shown in Figure 7.17. Name the project to be displayed and read the equivalent stress.

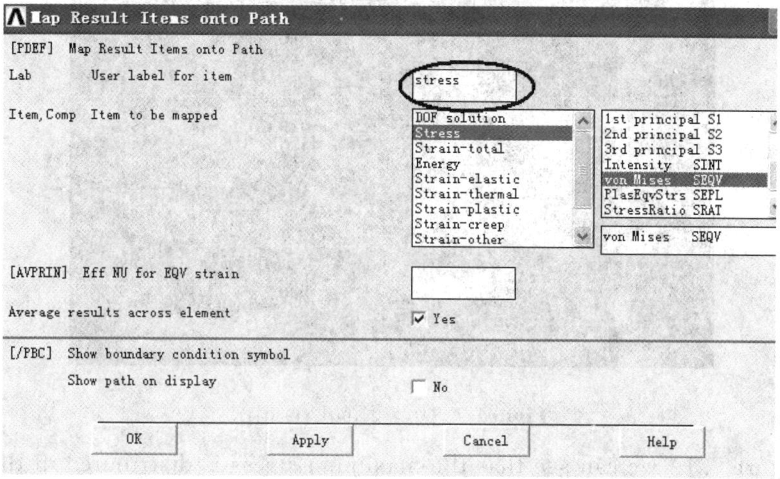

Figure 7.17 Parameter-read interface

Step 4: select the items (stress, displacement etc) to display on the path.
General Postproc→Path Operation→Plot Path Item→On Graph, select the corresponding data in accordance with the needs (shown in Figure 7.18), then click OK and get the results shown in Figure 7.19.

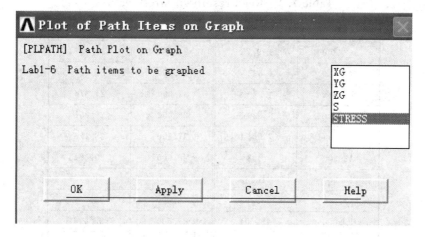

Figure 7.18 Read options

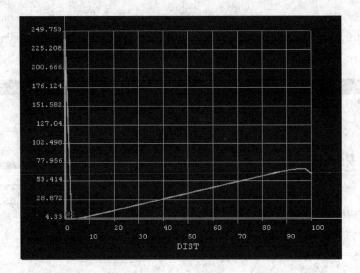

Figure 7.19 Read results

From Figure 7.19 we can see that the maximum stress is distributed in the end portion which is under the affection of concentrated force. Then, the stress is drastically reduced to a minimum point. And then gradually increase and come back to the maximum at the roots. To validate the results, we could use the method of listing data as follow:

General Postproc→Path Operation→Plot Path Item→List Path Items, select the defined display items (stress), get the stress chart of each point on the path shown in Table 7.2.

Table 7.2 Read results of stress values

S	S_T	S	S_T
0.000 0	249.75	27.500	20.627
2.500 0	4.622 7	30.000	22.501
5.000 0	4.330 2	32.500	24.376
7.500 0	5.830 5	35.000	26.251
10.000	7.581 0	37.500	28.126
12.500	9.408 4	40.000	30.001
15.000	11.265	42.500	31.876
17.500	13.132	45.000	33.751
20.000	15.004	47.500	35.626
22.500	16.878	50.000	37.501
25.000	18.752	52.500	39.376

(Continued)

S	S_T	S	S_T
55.000	41.251	80.000	59.982
57.500	43.126	82.500	61.833
60.000	45.001	85.000	63.652
62.500	46.876	87.500	65.399
65.000	48.750	90.000	66.985
67.500	50.625	92.500	68.248
70.000	52.500	95.000	68.994
72.500	54.374	97.500	69.286
75.000	56.247	100.00	62.764
77.500	58.118	—	—

In Table 7.2, S represents the distance and S_T represents the stress. Table 7.2 shows that the maximum stress is at the applying point of concentrated force. Then, the stress suddenly decreased. Subsequently, as gradually approaching the root, the stress value is gradually increased and reached the maximum at the root.

7.4.1.2 Simulation of stress distribution for two kinds of composite materials

In many cases, MEMS beams are not made of the same material, but made of two or more kinds of materials, while the excess material will have a great impact on performance. Because the thermal expansion coefficients of the two materials are different, when the temperature changes, there the internal stress will generate and make cantilever bend. But in many other sensors, the effect is easy to overlook leading to the calculation in accordance with a single material cantilever. Therefore, the calculation of composite materials is very important. However, there is a big difference in the finite element analysis of single-clamped beam between single material and composite materials. Firstly, you need to distribute different material attributes; Secondly, stick the two surfaces tightly, generally by using the command "glue", in order to make the two kinds of materials have the same displacement at each point at the interface.

1. Operation samples

Figure 7.20 shows a MEMS temperature sensor with single-clamped beam structure based on silicon substrate, and one aluminium metal layer is deposited on the surface of monocrystalline silicon cantilever. The bottom monocrystalline silicon beam dimensions of length 100 μm, width 10 μm, and thickness 4 μm. The aluminium layer dimensions are: length 100 μm, width 10 μm, thickness 2 μm. The cantilever is fixed at left.

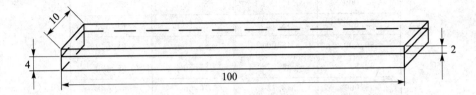

Figure 7.20 Calculation model (unit: μm)

Simulation: When the sensor is working at 50 ℃, the right-end of the beam will generate the displacement.

Material characteristics parameters are as follows.

Monocrystalline silicon: The elasticity modulus is 169 GPa, the Poisson's ration is 0.25, the density is 2 330 kg/m^3, the thermal expansion coefficient is 2.6×10^{-6}.

Aluminium: The elasticity modulus is 70 GPa, the Poisson's ration is 0.33, the density is 2 700 kg/m^3, the thermal expansion coefficient is 25×10^{-6}.

Computation method: This is a typical example of thermal-structure coupling analysis, and can be completed with Solid5 heat-structure coupling unit.

2. Calculation steps

(1) Preprocessing

● Add the element type:

Preprocessor→Element Type → Add/Edit/Delete→ Add→ Coupled Field→ Scalar Brick 5→ Close (set it as Solid5 unit, as shown in Figure 7.21).

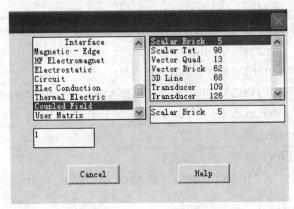

Figure 7.21 Add unit type

● Add material properties:

Because the sample is constituted with two kinds of materials, we need to add two kinds of material attributes. The method is Preprocessor→Material Props→Material Models.

Enter the interface shown in Figure 7.22.

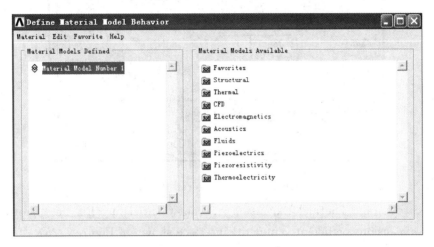

Figure 7.22 Addition of material attributes

Select: Structural → Linear → Elastic → Isotropic, to enter the material properties interface, fill in the elastic modulus and the Poisson's ratio corresponding to Micron unit (shown in Figure 7.23).

Select: Structural → Density, to add the relevant density (shown in Figure 7.24, corresponding to Micron unit).

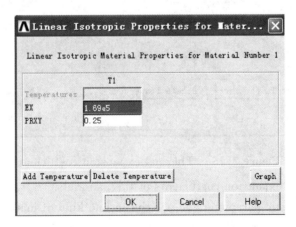

Figure 7.23 Material attributes

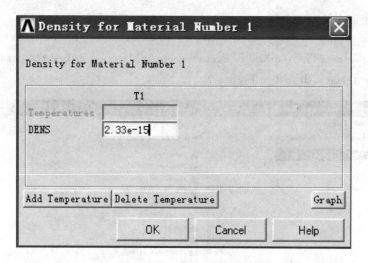

Figure 7.24 Material density

Select: Structural→Thermal Expansion→Secant Coefficient→Elastic→Isotropic, to add the thermal expansion coefficient (enter the interface shown in Figure 7.25).

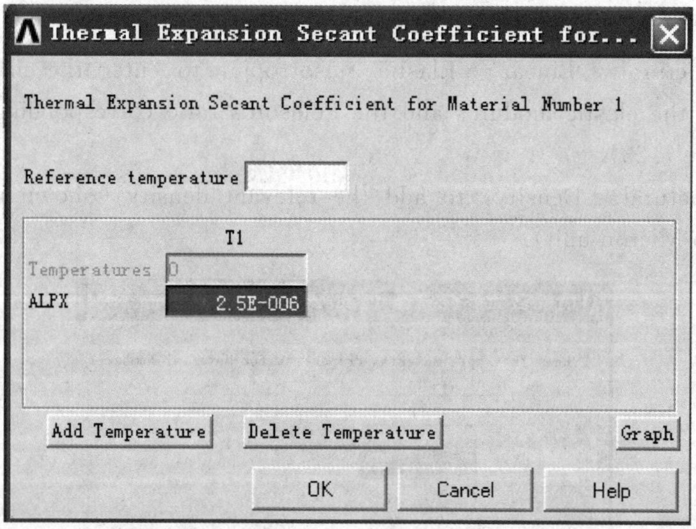

Figure 7.25 Thermal expansion coefficient

Fill in the thermal expansion coefficients of the materials.

Add the characteristic parameters of Al (the second kind of material).

Select: Edit→Copy, enter the interface shown in Figure 7.26.

Chapter 7 Application examples of the finite element method in the design of MEMS devices

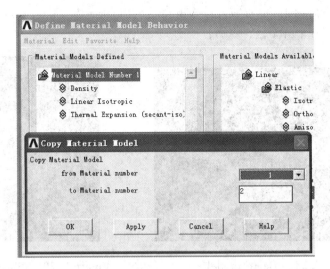

Figure 7.26 Copy material model

Set material parameters to 2, and then click characteristic parameters of the second kind of material, set the density, the elastic modulus, the Poisson's ratio and the thermal expansion coefficient parameters and all of the parameters of aluminium materials.

● Modelling:

Preprocessor → Modelling → Create → Volumes → Block → By Dimension, fill in the dimensions shown in Figure 7.27.

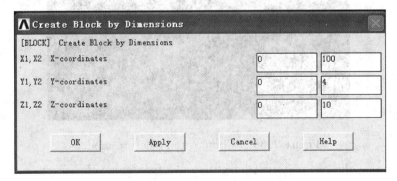

Figure 7.27 Dimensions of the beam

The model is shown in Figure 7.28:

Select: WorkPlane→Offset WP to→Keypoints, to shift the keypoints to the upper left corner of rectangle (shown in Figure 7.29).

● Draw aluminium structural layer:

239

Preprocessor→Modelling→Create→Volumes→Block→By Dimension.
Input (0,0,0) and (100,2,-10);

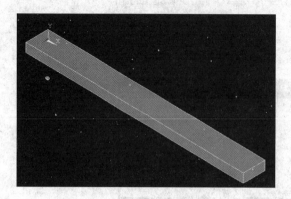

    Figure 7.28   Beam model   **Figure 7.29**   Shift the work plane

Select: PlotCtrls→Numbering, and select "Colors only" in "Volume number" to make the colors of two structures different(shown in Figure 7.30).

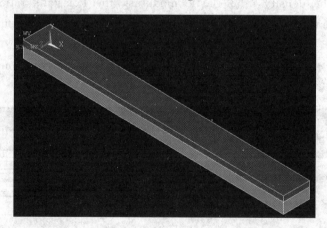

Figure 7.30   Composite structure model

To ensure that the two materials have the same freedom at the interface, select the Glue Command:

Preprocessor→Modelling→Operate→Booleans→Glue→Volumes, select "Pick All" after the glue interface pop-up.

- Meshing:

Select: Preprocessor→Meshing→Mesh Tool, to enter the right interface.

Select: Global→Set→Element Edge Length, to input the value 0.5.
- Material Attribute Assignment:

Because of those two different kinds of materials, their characteristic parameters are different. The software needs to assign the attributes of those two kinds of materials. Method: Select volumes→set in "Element Attributes" option, select monocrystalline silicon and click OK, then the interface (shown in Figure 7.31) pop-up, and click OK again.

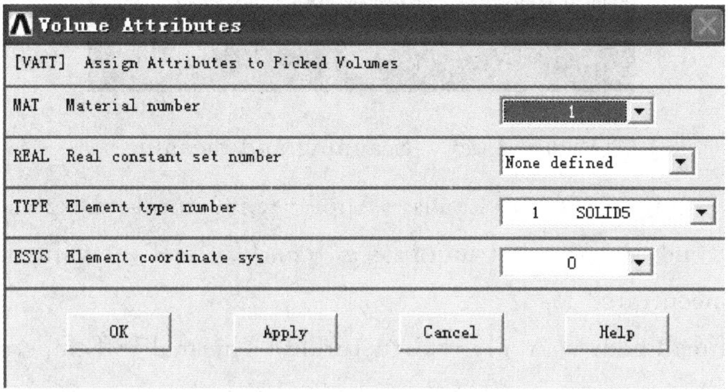

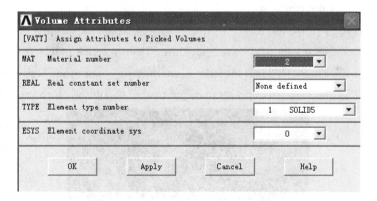

Figure 7.31   Silicon material attribute assignation

Similarly, select aluminium, set the material attribute as 2, then click OK(shown in Figure 7.32).

Figure 7.32   Al material attribute assignation

Select: Shape→Hex→Sweep, scan and mesh, as shown in Figure 7.33.

(2) Solution
- Set Analysis Type:

Solution→Analysis Type→New Analysis→Steady State.

Figure 7.33 Scanning and meshing

Define Loads: Solution→Define Loads→Apply→Structural→Displacement→On Areas, select the left-most end, set the restriction as x, y and z, and the value is 0.

● Add the concentrated load:

Solution→Define Loads→Apply→Structural→Thermal→Temperature→On Nodes, select "Pick All", then the interface pop-up, and set the temperature as 50 ℃.

● Calculation:

Solution→Solve→Current LS.

(3) General Postproc:

● **View the stress distribution:**

General Postproc→Plot Results→Contour Plot→Nodal Solution, enter the following interface, select the equivalent stress calculated item, click OK and get the stress distribution shown in Figure 7.34. It can be seen that the maximum stress does not exceed the allowable stress of the material, and its maximum value should be on the contact surface.

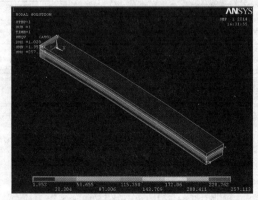

Figure 7.34 Stress distribution

Chapter 7 Application examples of the finite element method in the design of MEMS devices

From the distribution of the beam we know that its maximum stress is 249 MPa and is at the concentrated force point.

● View the figure of beam deformation under the concentrated force:

General Postproc→Plot Results→Deformed Shape, select Def+Undef Edge, click OK and get the deformed beam shown in Figure 7.35. The maximum value is 1.028 μm. Now it is clear that the sensor is sensitive to temperature.

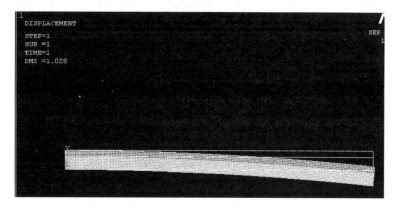

Figure 7.35　Deformed shape

● Stress distribution:

General Postproc→List Results→Contour Plot→Nodal Solution, then select the stress data.

● Determine the maximum displacement using path operations, and provide support for the sensor placement.

Step 1: Establish path.

General Postproc→Path Operation→Define Path→By Nodes, select the midpoint of both ends of the upper surface.

Get the following interface (shown in Figure 7.36) after select path and name the path "Guo".

Figure 7.36　Path selecting

Step 2: Call path.

There are many paths and we can call one of them. The specific operations are General

Postproc→Path Operation→Recall Paths, and select the required path as Figure 7.37 shows.

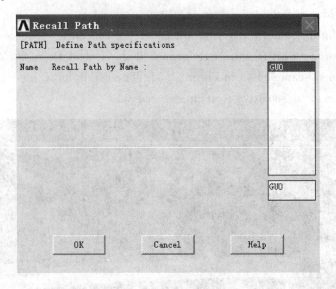

Figure 7.37 Path calling

Step 3: Select the items to display (stress, displacement etc.) on the path.

General Postproc→Path Operation→Map Onto Path, select item to be display as y-direction displacement, then name it "SS".

Step 4: Select the items to display (stress, displacement etc.) on the path.

General Postproc→Path Operation→Plot Path Item→On Graph, select corresponding displacement data according to the demand, click OK and get Figure 7.38.

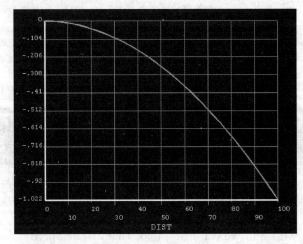

Figure 7.38 Curve of displacement results

Chapter 7 Application examples of the finite element method in the design of MEMS devices

7.4.2 Modal analysis of double-clamped beam

Double-clamped beam is often used in MEMS device. One of its uses is the supporting structure of the device, and another use is the sensing element of the double-clamped structure category, such as resonant tuning fork (shown in Figure 7.39) and stress sensitive beams assembly on the diaphragm of the pressure sensor which is used for sensing stress changes (shown in Figure 7.40). For double-clamped beam as the sensing element, its modal analysis is particularly important.

Figure 7.39 Double-clamped tuning-fork structure

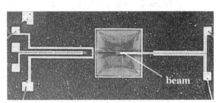

Figure 7.40 Sensitive beam of resonant pressure sensor

This subsection analyzes the modal of double-clamped beam under the load (such as the analysis of accelerometer resonance tuning-fork) and unloads two cases, in which the load can be divided into the centralized power load and the thermal load.

7.4.2.1 Natural resonant frequency analysis of double-clamped beam

1. Examples

A double-clamped beam is shown in Figure 7.41 with its length $L = 100$ μm, width $W = 60$ μm, thickness $H = 10$ μm. The material of the beam is silicon. The parameters can be seen in previous examples. The left side of the beam is constrained while the right side can only move axially along the beam, and we can calculate natural resonant frequency of the beam.

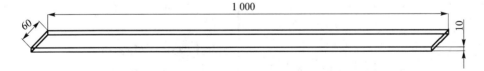

Figure 7.41 Parameters of double-clamped beam (unit: μm)

2. Analysis steps

(1) Preprocessor

● Add element type:

Preprocessor→Element Type→Add/Edit/Delete→Add→Solid→Brick 8 Node 45→Close, add Solid45.

● Add material properties:

Preprocessor→Material Props→Material Models, enter the interface shown in Figure 7.42.

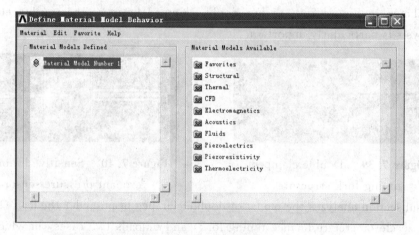

Figure 7.42 Add a unit type

● Select:

Structural→Linear→Elastic→Isotropic. Enter material properties, then add the elastic modulus and the Poisson's ratio corresponding to micron units (shown in Figure 7.43). After that, select Structural→Density, and add the density information (shown in Figure 7.44).

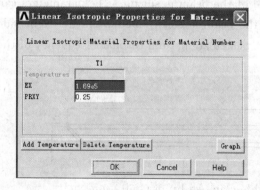

Figure 7.43 Material properties

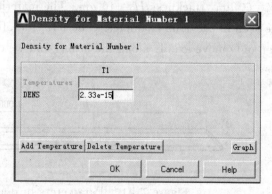

Figure 7.44 Add the density information

Chapter 7 Application examples of the finite element method in the design of MEMS devices

● Modelling:

Preprocessor → Modelling → Create → Volumes → Block → By Dimension, then add coordinate (0, 0, 0) and (1 000, 10, 16), and the model is shown in Figure 7.45.

● Meshing:

Preprocessor→Meshing→Mesh Tool, enter meshing operator interface. Select Global→Set→Element Edge Length, and add value 2; then select Shape→Hex→Sweep, and the sweeping generated grids are shown in Figure 7.46.

Figure 7.45 Beam model **Figure 7.46** Sweeping

(2) Solution:

● Set the type of analysis:

Solution→Analysis Type→New Analysis→Model.

● Set modal analysis options:

Solution→Analysis Type→Analysis Options, then enter the interface shown in Figure 7.47, and set the order as 3.

● Impose constraints:

Solution→Define Loads→Apply→Structrual→Displacement →On Areas, select the left and the right two ends. Select all DOFs, then add the value 0.

● Calculation:

Solution→Solve→Current LS.

(3) General Postproc:

● View the third-order resonance frequency:

General Postproc→Result Summary. Get the results shown in Figure 7.48.

● Read the first-order mode:

General Postproc→Read Results→First Set, Plot Ctrls→Animate→Mode Shape, get the interface shown in Figure 7.49, and it has personalized settings according to the requirements.

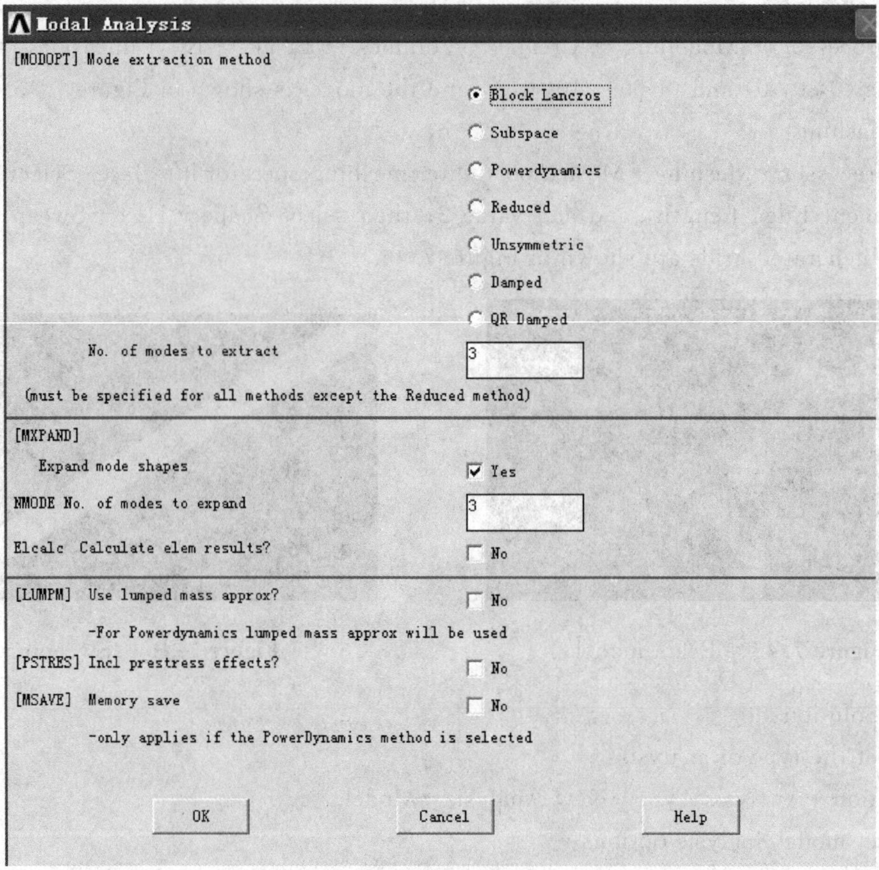

Figure 7.47 Selection of modal analysis parameters

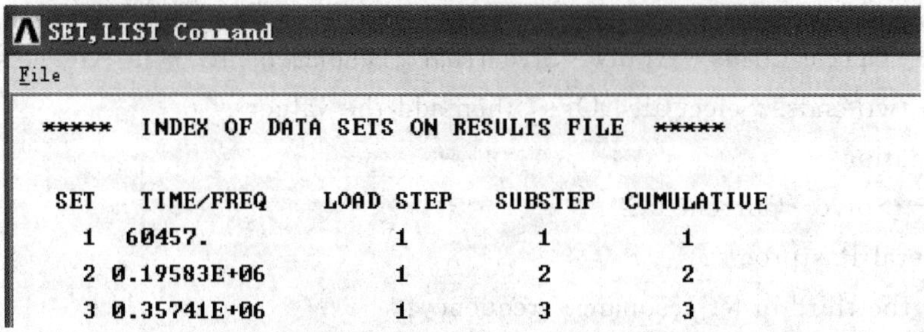

Figure 7.48 Results of modal analysis

Resonant type of the first-order mode is shown in Figure 7.50.

General Postproc→Read Results→Next Set, get the second-order mode shown in Figure 7.51.

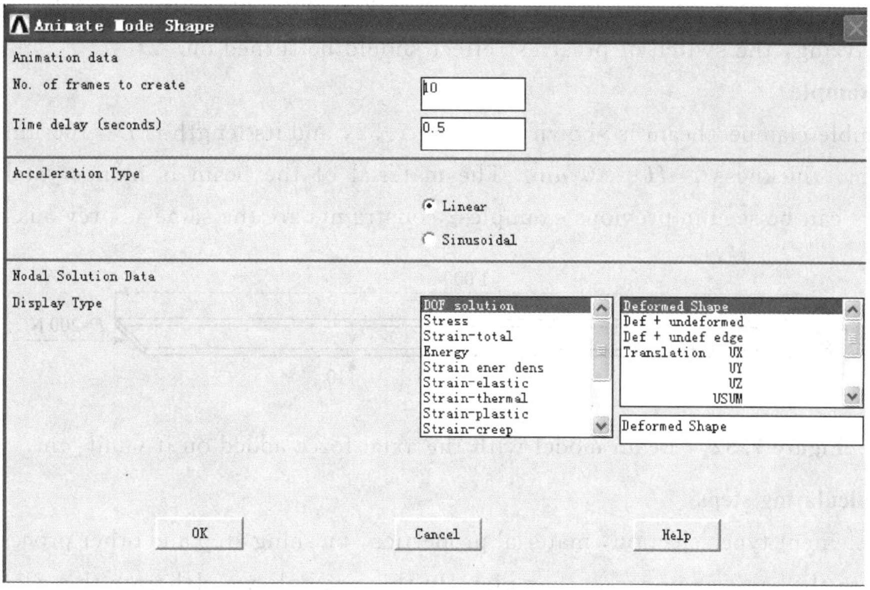

Chapter 7 Application examples of the finite element method in the design of MEMS devices

Figure 7.49 Demonstration of result parameter settings

Figure 7.50 The first-order mode

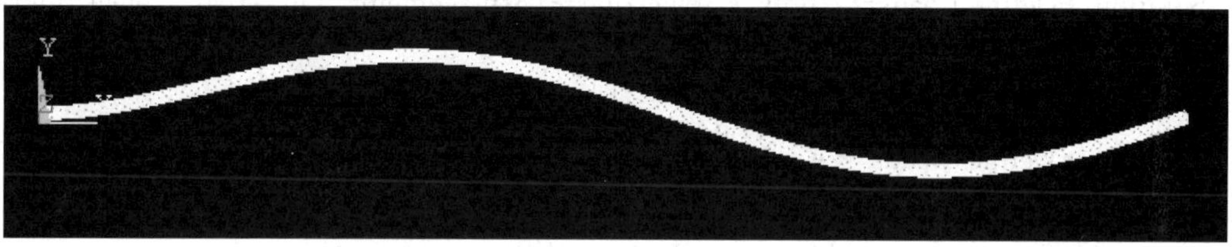

Figure 7.51 The second-order mode

7.4.2.2 Analysis of the resonant frequency after applying an axial force

This case is mainly applied to the simulation of the resonant frequency change when the load is applied to the sensor, which is similar to the modal analysis. But in calculation, static analysis goes first. In static analysis, you should open Include PreStress Effect option in

249

Solution Control interface. Then, do modal analysis without exiting calculation results. When analyzing, the switch of prestress effect should be turned on.

1. Example

A double-clamped beam is shown in Figure 7.52, and its length is $L=100$ μm, width is $W=60$ μm, thickness is $H=10$ μm. The material of the beam is monocrystalline. The parameters can be seen in previous examples; constraints are the same as previous examples.

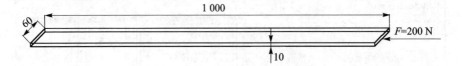

Figure 7.52 Beam model with the axial force added on it (unit: μm)

2. Calculating steps

The element type selection, material properties, meshing step and other processes of the example are the same as previous examples. In this example we elaborate the calculations of the solution process.

(1) Solution

● Set the type of analysis:

Solution→Analysis Type→New Analysis→Static.

● Add constraints, left end fully constrained.

Solution→Define Loads→Apply→Structrual→Displacement→On Areas, select the left end, then select all DOFs, and add value 0.

Solution→Define Loads→Apply→Structrual→Displacement→On Areas, select the right end, and set UY and UZ as 0.

● Add load.

Solution→Define Loads→Apply→Structrual→Force/moment→On Nodes, select the center of the right end, add value 200.

● Open Sol'n Control option:

Solution→Analysis Type→Sol'n Control, enter the interface as Figure 7.53 shows, and open Calculate PreStress Effect option in Basic target, click OK.

● Calculation:

After the calculation is completed, do not check any results, and do modal calculation.

① Solution→Analysis Type→New Analysis→Mode.

② Set modal analysis options: Solution→Analysis Type→Analysis Options. Type 3 in

Chapter 7 Application examples of the finite element method in the design of MEMS devices

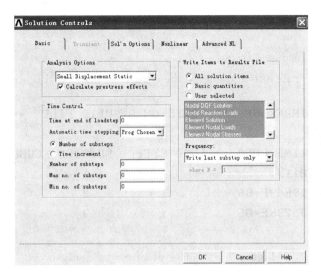

Figure 7.53 PreStress option

the interface table, Then open Incl PreStress Effect option in the interface (shown in Figure 7.54), and the frequency range can be set as 0. Then we can analyze any frequency.

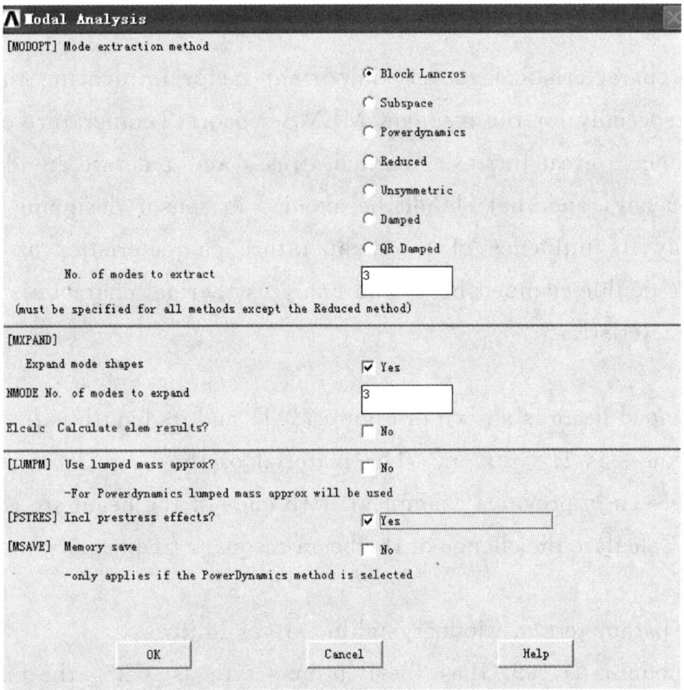

Figure 7.54 PreStress option opening

③ Calculation: Solution→Solve→Current LS.

(2) General Postproc

● View 3 order resonant frequency:

General Postproc→Result Summary, the result is shown in Figure 7.55.

```
***** INDEX OF DATA SETS ON RESULTS FILE *****

  SET   TIME/FREQ    LOAD STEP   SUBSTEP   CUMULATIVE
   1    60113.           1          1          1
   2    0.19543E+06      1          2          2
   3    0.35735E+06      1          3          3
```

Figure 7.55 Result of modal analysis

From the analysis result, when 200 μN axial force is loaded, the first-order resonant frequency changes from 60 457 Hz to 60 113 Hz. The change is obvious and illustrates the correction of analytical methods.

7.4.2.3 Impact of heat load resonant beam analysis

Temperature characteristic is a very important factor influencing the performance of MEMS devices, especially for the resonant MEMS sensor. Temperature changes will cause the sensor to produce a great intrinsic thermal stress, and generate an additional change in the resonant frequency, and that should be avoided in sensor designing. Therefore, it is important to study its influence of the temperature characteristics of the device. This example select the double-clamped beam and analyzes thermal characteristics' influences on its frequency characteristics.

1. Example

A double-clamped beam is shown in Figure 7.41, and its length is $L=100$ μm, width is $W=60$ μm, thickness is $H=10$ μm. The material of the beam is monocrystalline. The parameters can be seen in previous examples; both ends of the beam are constraint.

Calculation: Calculate the change of the beam resonant frequency when 10 ℃ is added to the beam.

The material parameters of Monocrystalline are as follows:

The elastic modulus is 169 GPa, the Poisson's ratio is 0.25, the material density is 2 330 kg/m^3, and the thermal expansion coefficient is 2.5×10^{-6}.

2. Calculation process

(1) Preprocessor

● Add element type:

Preprocessor→Element Type→Add/Edit/Delete→Add→Coupled Field→Scalar Brick 5→Close, add Solid5 unit.

● Add material properties as the previous operation process (refer to Subsection 7.4.1.1).

● Modelling:

Preprocessor→Modelling→Create→Volumes→Block→By Dimension.

Add coordinate (0,0,0) and (1 000,10,60), the model is shown in Figure 7.56.

Figure 7.56 Finite element model of the beam

● Meshing:

Preprocessor→Meshing→Mesh Tool, then enter the grid operation interface.

Select Global→Set→Element Edge Length, and add value 2;

Select the straight line length as 1 000 and set the subdivision as 200 in equality; select the straight line length as 60 and set the subdivision as 20 in equality.

Select Shape→Hex→Sweep, and sweep generated grids as Figure 7.57 shows.

Figure 7.57 Finite element of grids

(2) Solution
- Set analysis type:

Solution→Analysis Type→New Analysis→Model.
- Set modal analysis type:

Solution→Analysis Type→Analysis Options, then enter the interface as Figure 7.58 shows, and set order as 3.

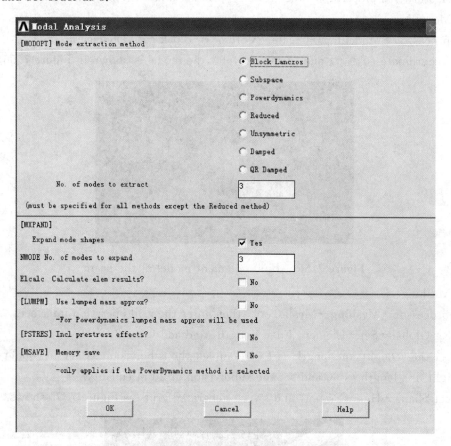

Figure 7.58 Selection of modal order

- Add constraints:

Solution→Define Loads→Apply→Structrual→Displacement→On Areas, select the left and the right two end faces, and select all DOFs, then add value 0.
- Calculation:

Solution→Solve→Current LS.

(3) General Postproc (a)

- View 3-order resonant frequency:

General Postproc→Result Summary, and get results as Figure 7.59 shows. Referring to previous examples, we find the beam resonant frequency changes due to its different ways of constraints.

Calculating resonant frequency changes after temperature load is applied.

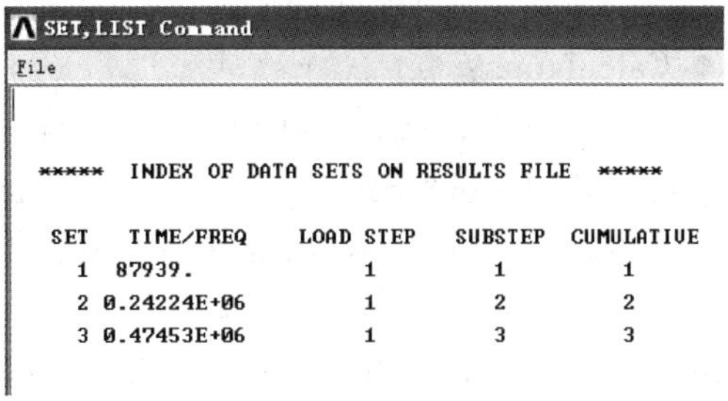

- Set analysis type:

Solution→Analysis Type→New Analysis→Static.

- Add constraints, fully constrained.
- Add load:

Solution→Define Loads→Apply→Thermal→Temperature→On Nodes, select "Pick All", and set value as 10.

- Open Sol'n Control option:

Solution→Analysis Type→Sol'n Control, and enter the interface as Figure 7.60 shows.

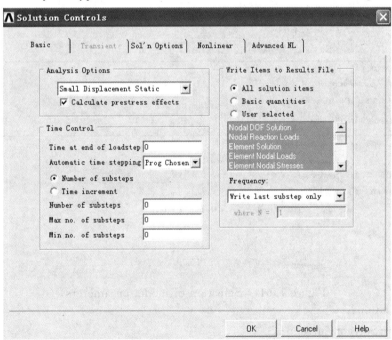

Figure 7.59 Modal analysis results (a)

Figure 7.60 Open PreStress option

open Calculate PreStress effects option in Basic label, then click OK.
- Calculation:

Solution→Solve→Current LS.

After calculation, continue the modal calculation without checking the result.
- Solution→Analysis Type→New Analysis→Mode.
- Set modal analysis options:

Solution→Analysis Type→Analysis Options, in order to set interface, set order as 3. Then open Incl PreStress effect option in the interface as Figure 7.61 shows. The frequency range can be set as 0, thus we can analyze situations under any frequency.

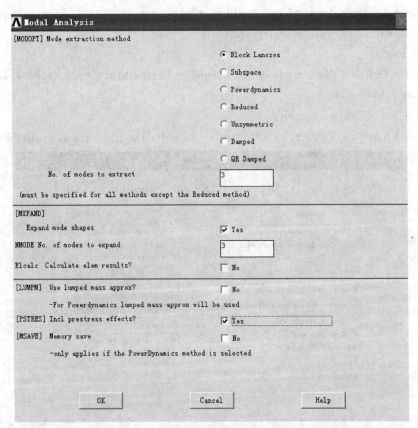

Figure 7.61 Selection of modal parameters

- Calculation:

Solution→Solve→Current LS.

(4) General Postproc (b)
- View 3-order resonant frequency:

General Postproc→Result Summary, and get the following results as Figure 7.62 shows.

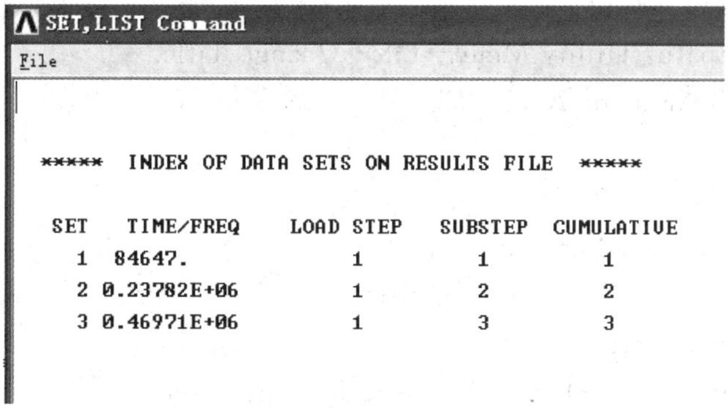

```
***** INDEX OF DATA SETS ON RESULTS FILE *****

SET    TIME/FREQ    LOAD STEP    SUBSTEP    CUMULATIVE
 1     84647.           1            1           1
 2     0.23782E+06      1            2           2
 3     0.46971E+06      1            3           3
```

Figure 7.62 Modal analysis results (b)

It is clear that the change of temperature has a great impact on the resonant frequency of double-clamped beam, in device designing process we should try to avoid the full-clamped situation at both ends.

7.4.3 Capacitance analysis of MEMS electrostatic comb fingers drive

Besides the double-clamped beam, MEMS electrostatic comb fingers drive is another very important kind of driver source in MEMS. The electrostatic actuation-capacitance detection method has become the first choice of almost all MEMS devices because of its advantages of highly integrated, easy to assemble, etc. Currently, the capacitance calculation of most such devices is based on the ideal equation. But from the analysis in Chapter 2, the theory has a large error. Therefore, after the design finished, using the finite element method to do simulation has become a very important research section.

An operation example is introduced below.

As shown in Figure 7.63, there are 90 pairs of electrostatic comb fingers drive. The dimensions of the comb fingers are: width 4 μm, length 100 μm, thickness 72 μm, gap of the comb fingers 4.5 μm, overlap length 50 μm. Then add potential difference of 5 V load, and calculate comb capacitance value.

Figure 7.63 Comb finger calculating model

Step 1: Initialization.
① Enter Ansys program.
② Select menu path: Utility Menu→File→Change Title.
③ Enter "Comb Actuator Analysis".
④ Click OK.
⑤ Select Main Menu→Preferences.
⑥ Click Magnetic-Nodal and Electric.
⑦ Click OK.

Step 2: Define parameters.
① Select: Utility Menu→Parameters→Scalar Parameters.
② Add the following parameters (if an input error occurs, please re-enter):
   $V1=5.0$;
   $V0=0.0$.
③ Click Close.

Step 3: Define cell types.
① Select Main Menu→Preprocessor→Element Type→Add/Edit/Delete.
② Click Add.
③ Click highlighted "Electrostatic" and "3D Tet 123".
④ Click OK.
⑤ Click Close.

Step 4: Define material properties.
① Select: Main Menu→Preprocessor→Material Props→Material Models.
② In material window, double-click the following options: Electromagnetics, Relative Permittivity, Constant.
③ PERX (Relative Permittivity) input 1, click OK. The material number displayed in the left area of the material defined window is 1.
④ Select menu path: Edit→Copy. Click OK. Copy material 1 to material 2.
⑤ In material box, double-click No. 2 material and Permittivity (constant).
⑥ Enter 11.5 in PERX area, click OK.
⑦ Select menu path: Material→Exit.
⑧ Click SAVE_DB on the Ansys Toolbar.

Step 5: Establish the geometry.
Establish comb fingers model:
① Select: Main Menu → Preprocessor → -Modeling-Create → -Volumes-Block → By

Chapter 7 Application examples of the finite element method in the design of MEMS devices

Dimensions.

② Enter the following values:

−2.25	−6.25
−25	75
−36	36

③ Click Apply.

④ Enter the following values:

2.25	6.25
−75	25
−36	36

⑤ Click OK.

⑥ Copy the entity: Main Menu→Preprocessor→-Modeling-Copy→-Volumes.

⑦ Click Pick All.

⑧ After DX X-offset in active CS enter 17.

⑨ Click Apply.

⑩ Click Pick All.

⑪ After DX X-offset in active CS enter 34.

⑫ Click Apply.

⑬ Click and select one pair of comb fingers at the rightmost.

⑭ Click OK.

⑮ After DX X-offset in active CS enter 17.

⑯ Click Apply.

⑰ Change select method from Single to Box.

⑱ Circle left four pairs of comb.

⑲ Click OK.

⑳ After DX X-offset in active CS enter −68.

㉑ Click OK.

㉒ Click SAVE_DB on the Ansys Toolbar.

㉓ Select: Main Menu → Preprocessor → -Modeling-Create → -Volumes-Block → By Dimensions.

㉔ Enter the following values:

−65.75	74.25
−75	−108
−36	36

㉕ Click Apply.

㉖ Enter the following values:

65.75	−74.25
75	108
−36	36

㉗ Click OK.

㉘ Add comb and beam together: Main Menu→Preprocessor→-Modeling-Operate→-Booleans-Add→Volumes.

㉙ Select 9 top comb fingers and top beam.

㉚ Click Apply.

㉛ Similarly, add 9 bottom comb fingers and bottom beam together.

㉜ Select Utility Menu→Plot Ctrls→Pan, Zoom, Rotate.

㉝ Click Iso button, click Close.

Establish air model:

㉞ Select Main Menu→Preprocessor→-Modeling-Create→-Volumes-Block →By Dimensions.

㉟ Enter the following values:

−150	150
−200	200
−100	100

㊱ Click OK.

㊲ Main Menu→Preprocessor→-Modeling-Operate→-Booleans-Overlap→Volumes.

㊳ Click Pick All.

㊴ Select Utility Menu→Plot→Volumes.

㊵ Select Main Menu→Finish.

Step 6: Grid division and material settings.

① Select: Main Menu→Preprocessor→-Meshing-MeshTool.

Chapter 7 Application examples of the finite element method in the design of MEMS devices

② Click Set in Element Attributes.

③ Select 2 in drop-down menu of Material number.

④ Click OK.

⑤ Tick Smart Size.

⑥ Move Smart Sizing slider to 3.

⑦ Click MESH button.

⑧ Click on the upper and lower parts of the comb.

⑨ Click OK.

⑩ Select: Main Menu→Preprocessor→-Meshing-MeshTool.

⑪ Click Set in Element Attributes.

⑫ Select 1 in drop-down menu of Material number.

⑬ Click OK.

⑭ Click MESH button.

⑮ Click on air model.

⑯ Click OK.

⑰ Select: Main Menu→Preprocessor→-Meshing-MeshTool.

⑱ Click Close.

⑲ Click SAVE_DB on the Ansys Toolbar.

Step 7: Applying boundary conditions and loads.

① Select: Utility Menu→Select→Entities.

② Set top button as "Volumes".

③ Click OK.

④ Select: Utility Menu→Select→Entities.

⑤ Click OK.

⑥ Select: Utility Menu→Select→Entities.

⑦ Set top button as "Nodes".

⑧ Set lower button as "Attached to", then select Volumes→All.

⑨ Click Apply.

⑩ Click Plot.

⑪ Select: Main Menu → Preprocessor → Loads → Define Loads → -Apply-Electric → -Boundary→-Voltage-On Nodes.

⑫ Click Pick All.

⑬ Enter V1 at "Value of voltage (VOLT)" zone.

⑭ Click OK.

⑮ Select: Utility Menu→Select→Entities.

⑯ Set top button as "Volumes".

⑰ Set lower button as "By Num/Pick".

⑱ Click OK.

⑲ Click on lower part of the comb.

⑳ Click OK.

㉑ Select: Utility Menu→Select→Entities.

㉒ Set top button as "Nodes".

㉓ Set lower button as "Attached to"→Volumes→All.

㉔ Click Apply.

㉕ Click Plot.

㉖ Select: Main Menu → Preprocessor → Loads → Define Loads → -Apply-Electric → -Boundary→-Voltage-On Nodes.

㉗ Click Pick All.

㉘ Enter V0 at "Value of voltage (VOLT)" zone.

㉙ Click OK.

Step 8: solving.

① Click OK.

② Select: Utility Menu→Select→Everything.

③ Click: Utility Menu→-Plot-Nodes.

④ Select: Main Menu→Solution→-Solve-Current LS.

⑤ After the confirming message click File→-Close.

⑥ Click OK start solving. After solving will pop up a message, click Close.

⑦ Select: Main Menu→Finish.

Step 9: Saving analysis results.

① Select: Main Menu→General Postproc→Element Table→Define Table.

② Click Add.

③ Enter SENE in "User label for item" zone.

④ Light "Energy" in "Results data item" zone. (when the left "Energy" displays in high brightness, the right "Elec energy SENE" automatically appears in high brightness)

⑤ Click OK.

⑥ Click Add.

⑦ Enter EFX in "User label for item" zone.

⑧ Light "Flux & gradient" and "Elecfield EFX" in "Results data item" zone.

Chapter 7 Application examples of the finite element method in the design of MEMS devices

⑨ Click OK.
⑩ Click Add.
⑪ Enter EFY in "Results data item" zone.
⑫ Light "Flux & gradient" and "Elec field EFY" in "Results data item" zone.
⑬ Click OK.
⑭ Click Add.
⑮ Enter EFY in "User label for item" zone.
⑯ Light "Flux & gradient"and "Elec field EFY" in "Results data item" zone.
⑰ Click OK.
⑱ Click Close.
⑲ Click SAVE_DB.

Step 10: Drawing results figure.
① Select: Main Menu→General Postproc→Plot Results→-Vector Plot-Predefined.
② Click OK.

Step 11: Capacitance calculation.
① Select: Main Menu→General Postproc→Element Table→Sum of Each Item.
② Click OK. A pop-up window will show a chart of all units and their values.
③ Click Close.
④ Select: Utility Menu→Parameters→Get Scalar Data.
⑤ Light "Results data" and "Elem table sums" in "Type of data to be retrieved" zone.
⑥ Click OK. A pop-up dialogue will show the value of added unit chart.
⑦ Enter W in "Name of parameter to be defined" zone.
⑧ Set "Element table item" zone as "SENE".
⑨ Click OK.
⑩ Select: Utility Menu→Parameters→Scalar Parameters.
⑪ Add the following values:
$$C = (w*2)/[(V1-V0)**2]$$
$$C = [(C*10)*1e6]$$
⑫ Click Close.
⑬ Select Utility Menu→List→Status→Parameters→Named Parameter.
⑭ Light C in "Name of parameter" zone.
⑮ Click OK. A pop-up window will show the value of C.
⑯ Click OK to close the pop-up window.

7.4.4 Fatigue strength calculation example

7.4.4.1 Concept of fatigue

Fatigue is the phenomenon of structure fatigue fracture under repeatedly alternating load below its static limit load. For example, a pole can withstand 300 kN static tensile, but can be broken after through a certain number of cycles. Factors affecting fatigue contains alternating load cycles experienced, stress amplitude, mean stress, and local stress concentration etc.

7.4.4.2 Principle of fatigue analysis

Fatigue analysis of Ansys uses classic Miner linear cumulative damage theory.

If the member is under a constant stress S and recycled to the destruction, hypothesize the fatigue life is N. Define the damage through the n-th cycle is $D=n/N$.

It is clear that for the member under the a constant stress S, if $n=N$, $D=1$, and the member will fatigue failure.

When the member is under the stress S_i, the damage of the n_i-th cycle is $D_i=n_i/N_i$. If it is under stress S_i and through n_i cycles, define its total damage as (the fatigue life is a coefficient in Ansys)

$$D = \sum_{i=1}^{k} D_i = \sum_{i=1}^{k} n_i/N_i \qquad (7.1)$$

Failure criterion is

$$D = \sum_{i=1}^{k} D_i = 1 \qquad (7.2)$$

This is the Miner linear cumulative damage theory. n_i is the number of cycles in the action of the stress S_i. N_i is the life of the member recycled to damage under the stress effect of S_i, which is determined by the S-N curve.

7.4.4.3 Ansys fatigue analysis example

1. Project example

Assess the life of MEMS plate capacitor under a certain pressure, its model equals to a rectangular plate as Figure 7.64 shows. The size of the rectangular plate is 8 mm×10 mm× 0.05 mm (width × height × thickness). The bottom of the plate is fixed; the freedom degree in every axis direction is 0. Applying 16 034 Pa uniform pressure on the rectangular plate, analyze fatigue conditions and load times can be loaded on the corresponding plate.

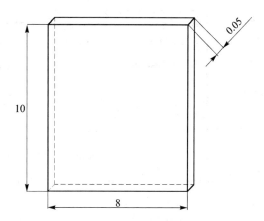

2. Brief analysis steps

Fatigue calculation is carried out in general postprocessor POST1, and should finish stress calculation before. Fatigue calculation usually contains five main steps as follows.

Figure 7.64 Model analysis

(1) Enter general postprocessor POST1, resume database.

(2) Set size, design the number of positions, and load event: Main Menu→General Postproc→Fatigue→Size Settings.

Define Fatigue material properties, input S-N curve: Main Menu→General Postproc→Fatigue→Property Table→SN Table.

Determine the stress position, define the stress concentration factor: Main Menu→General Postproc→Fatigue→Stress Locations.

(3) Extract and save the stress of different loads and events on the location of interest, specify the event loop and scale factor.

Extraction stress: Main Menu→General Postproc→Fatigue→Store Stresses→From rst File.

Save stress, specify the event loop and scale factor: Main Menu→General Postproc→Fatigue→Assign Events.

(4) Activate fatigue calculation: Main Menu→General Postproc→Fatigue→Calculate Fatigue.

(5) View the result.

3. Specific analysis

(1) Open Ansys, change the task name.

Pick up the menu: Utility Menu→File→Change Jobname, then pop-up a dialogue as Figure 7.65 shows.

Modify jobname, then click OK.

(2) Select preferences and unit types.

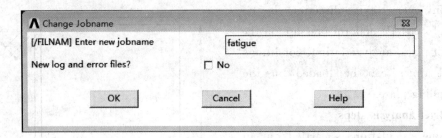

Figure 7.65 File name dialogue

Pick up the menu: Main Menu→Preferences→Select Structure.

Select unit type: Main Menu→Preprocessor→Element Type→Add/Edit/Delete. In pop-up dialogue as click Add button; pop-up the dialogue shown in Figure 7.66, the left list select "Solid", in the right list select "20node95", click OK, then click Close.

(3) Define material model.

Pick up the menu: Main Menu→Preprocessor→Material Props→Material Models. Enter the material parameters and the density as Figure 7.66 and Figure 7.67 show.

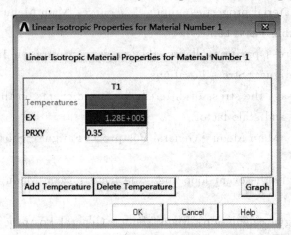

Figure 7.66 Material parameters

(4) Modelling: Utility Menu→Preprocessor→Modelling→Create→Volumes→Block→By Dimensions. Pop-up the window as Figure 7.68 shows, then enter the length, width and height of the cube, and click OK.

(5) Meshing: Preprocessor → Meshing → Mesh Tool, Set → Global → Size → Element Length, set universal size as 0.05 and it will be divided into grids as Figure 7.69 shows.

(6) Impose constraints.

Chapter 7 Application examples of the finite element method in the design of MEMS devices

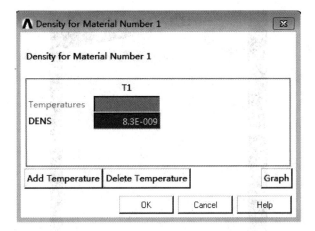

Figure 7.67 Density parameter inputting

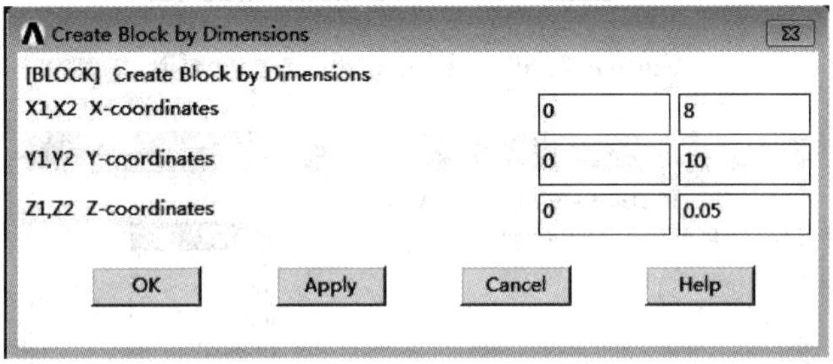

Figure 7.68 Model parameters inputting

Analysis unit settings: Utility Menu→Solution→Analysis Type→New Analysis→Static.

Impose constraints: Utility Menu→ Solution→ Define Loads→ Apply→ Structural→ Displacement → On areas, select the fixed upper and lower bottom surfaces.

After selection, click OK, we will get a window as Figure 7.70 shows.

Select "All DOF" in DOFs to be constrained. Enter 0 as the value, and click OK.

Add load: Utility Menu→Solution→Define Loads →Apply →Structural →Pressure → On Areas. In the pop-up window, select surface to apply uniform pressure, click OK.

In the pop-up window, enter uniform pressure values in Load PRES value, and click OK.

(7) Solution.

Select Utility Menu→Solution→Solve→Current LS.

The calculation window will pop-up, and click OK to start the calculation.

Until the calculation is over, click Close to quit.

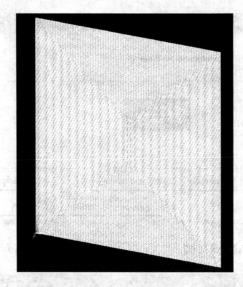

Figure 7.69 Finite element grids

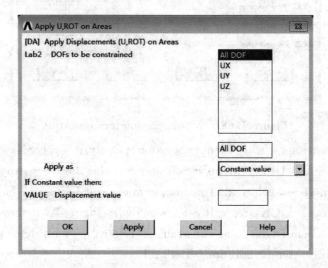

Figure 7.70 Adding constraints

(8) Size setting.

Pick up the menu: Main Menu→General Postproc→Fatigue→Size Settings.

The pop-up dialogue is shown in Figure 7.71. Enter 1 (position) in "MXLOC" textbox, enter 2 (case) in "MXEV", enter 2 (load) in "MXLOD", and click OK.

(9) Enter S-N curve.

Chapter 7 Application examples of the finite element method in the design of MEMS devices

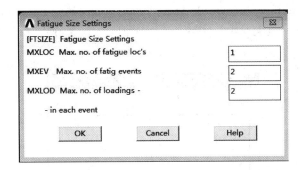

Figure 7.71 Fatigue settings

Main Menu→General Postproc→Fatigue→Property Table→S-N Table and the pop-up dialogue is shown in Figure 7.72, enter stress S and life N in chart. S-N curve differs in case of different materials; click OK after the values are entered.

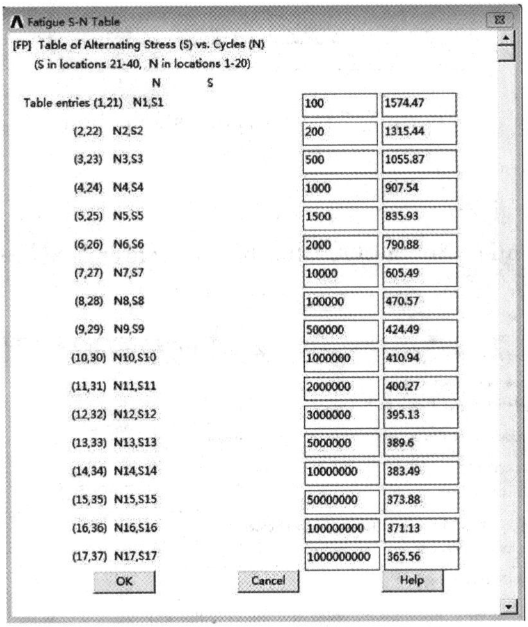

Figure 7.72 S-N curve

(10) Query node number and save it to variable.

Pick up the menu: Utility Menu→Parameters→Scalar Parameters, and the pop-up dialogue is shown in Figure 7.73. Enter NNN = NODE (0.1, 0.05, 0) in "Selection" textbox, click Accept and then click Close.

269

Search the serial number of coordinates (0.1, 0.05, 0) node, and save it to variable NNN. There is a quite big stress in the node, so there is a great possibility for the fatigue failure.

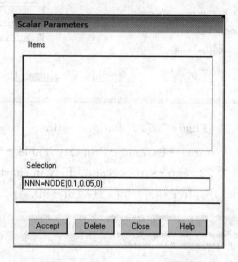

Figure 7.73 Keypoints selection

(11) Specify stress position.

Main Menu→General Postproc→Fatigue→Stress Locations. The pop-up dialogue is shown in Figure 7.74, enter 1 in "NLOC" textbox and enter NNN in "NODE" textbox, then click OK.

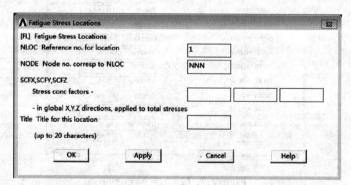

Figure 7.74 Position selection dialogue

(12) Extract stress of event 1.

Pick up the menu: Main Menu→General Postproc→Fatigue→Store Stresses→From rst File. The pop-up dialogue is shown in Figure 7.75. Enter NNN (position) in "NODE"

Chapter 7 Application examples of the finite element method in the design of MEMS devices

textbox, enter 1 (event) in "NEV" textbox, and enter 2 (load) in "NLOD" textbox, then click OK.

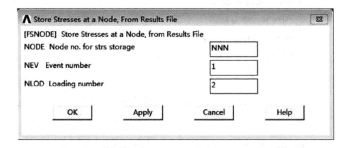

Figure 7.75 Stress access dialogue

(13) Save stress of event 1.

Pick up the menu: Main Menu→General postproc→Fatigue→Store Stresses →Specified Val. The pop-up dialogue is shown in Figure 7.76, enter NNN (position) in "NODE" textbox, enter 1 (event) in "NEV" textbox, enter 1 (load) in "NLOD" textbox, click OK. Then click OK in the pop-up dialogue.

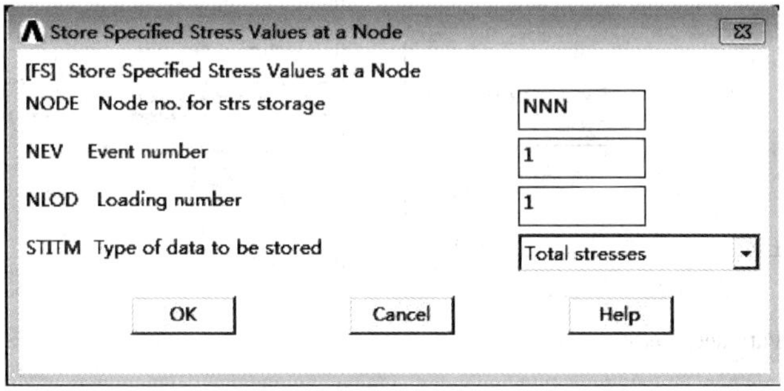

Figure 7.76 Stress access dialogue

(14) Specifies the event repetitions.

Pick up the menu: Main Menu→General Postproc→Fatigue→Assign Events. The pop-up dialogue is shown in Figure 7.77, enter 1 (event) in "NEV" textbox, enter 1 000 (repetitions) in "CYCLE" textbox, enter 1 (stress proportion) in FACT textbox, then click OK.

Repeat step (12) ~ step (14). When repeat step (12), enter 2 (event) in "NEV" textbox; when repeat step (13), enter 2 (event) in "NEV" textbox; when repeat step (14), enter 2 (event) in "NEV" textbox, enter 600 000 (repetitions) in "CYCLE" textbox, and enter 1.1 (stress proportion) in FACT textbox.

271

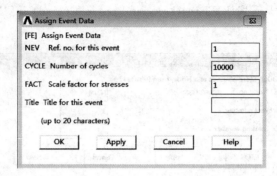

Figure 7.77 Repetitions selection dialogue

(15) Fatigue calculation

Pick up the menu: Main Menu → General Postproc → Fatigue → Calculate Fatigue. The pop-up dialogue is shown as follows, and then click OK. The calculated result is shown in Figure 7.78, and the fatigue life coefficient is 0.000 01.

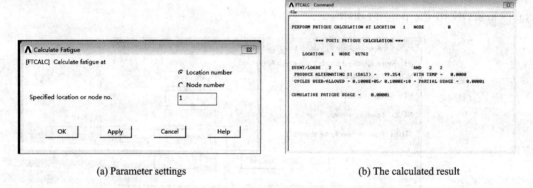

(a) Parameter settings (b) The calculated result

Figure 7.78 Calculation result of fatigue life

7.5 Summary

This chapter introduces the application of FEM method in the design of MEMS devices based on a large number of engineering examples in detail. The main targets of examples are the most commonly used single-clamped beams, the double-clamped beams and the comb capacitor drives in MEMS. Meanwhile, this chapter also introduces many operating skills, so as to provide some references to the readers.